PRAISE FOR GUARDIANS

"An enthralling up-close account... Intense... Brilliant... Unforgettable. Maki puts her foot on the accelerator and doesn't let go!"

— George Beck, Ph.D.
Author of Law and Enforcement in American Cinema, 1894-1952

"In a day & age where our Heroes are viewed as robots, vilified & insensitivity is the faux reality created through television and cell phone screens, Julia paints a vivid picture of true service. As the reader quickly becomes familiar with her willingness to express her vulnerability, she opens your mind to an experience that none of us would want to live through in these great United States.

"Her experiences are not much different from the stories I have heard from fellow brothers and sisters in arms who spent time in combat with their hearts set on protecting others, while watching those same people turn against them. It is heartbreaking knowing how the hate right here in our own country is still being spewed toward OUR Heroes regardless of sex or ethnicity. Julia shows the very human side to being on the receiving end of loving your country and fellow American, but being hated for it because you choose to wear a uniform."

— Ryan Weaver, CW3(R)
Former Blackhawk Aviator, Country Music Artist

GUARDIANS

Stories of the 2020 Civil Unrest
in Washington, D.C.

By Julia A. Maki Pyrah
with Nicholas A. Cotroneo

GUARDIANS

Stories of the 2020 Civil Unrest in Washington, D.C.

Copyright © 2020 by Julia A. Maki Pyrah

All rights reserved. No part of this book may be reproduced or utilized in any form or by any means, electronic or mechanical, including photocopying and recording, or by any information storage and retrieval system, without permission in writing from the publisher.

First Edition

Because of the dynamic nature of the internet, any web address or links contained in this book may have changed since publication and may no longer be valid.

The views expressed in this work are solely those of the author and do not necessarily reflect the views of the publisher, and the publisher hereby disclaims any responsibility for them.

Cover photo taken by: Julia Pyrah
Bio photo taken by: Barry Morgenstein

Published by Tactical 16, LLC
Monument, CO

ISBN: 978-1-943226-52-8 (paperback)

CONTENTS

All the Thanks v

Prologue 1

Early May 2020 3
An Inequality of Sorts

Monday, May 25, 2020 5
The Waves of Injustice

Tuesday, May 26, 2020 7
The Blowing Wind

Wednesday, May 27, 2020 9
Sparks Fly

Thursday, May 28, 2020 11
Fire Spreads

Friday, May 29, 2020 13
Reaction

January 2017 15

Saturday, May 30, 2020 17
Be Advised

Sunday, May 31, 2020 21
Ready, Set, Go!

March 2017 27

Monday, June 1, 2020 29
Boots on the Ground

May 2017 37

Photos 39

Early Hours of June 2nd, 2020 47

Tuesday, June 2nd, 2020 51
12th & Constitution

June 2017 61

Wednesday, June 3, 2020 63
Marley's World

Thursday, June 4, 2020 79
Riot Training at Ellipse Park

Friday, June 5, 2020 93
Escort Service

Saturday, June 6, 2020 107
Weekend Vibes

Tuesday, June 9, 2020 113
L Street & 11th

Wednesday, June 10, 2020 121
Day Off Down South

Thursday, June 11, 2020 125
Back to the Streets

Saturday, June 13, 2020 133
Armory/US Park Police Station

Sunday, June 14, 2020 137
The COVID

Tuesday, June 16, 2020 141
Leaving the War

Epilogue 147

A Note from the Author 151

About the Author

About the Co-Author

About the Publisher

ALL THE THANKS

Although you're about to read my story, the truth is, there are so many others that have played a substantial role in this story. Whether mentioned, personal accounts, or behind the scenes.

First, there was this fabulous crew that took the time to put together goodie bags of happiness for all of my airmen and soldiers. Let me tell you—the huge box was gone in minutes and it was definitely a highlight of our week. So thank you to the following: the Somers family (for coordinating and contributing), LT. CDR (Ret.) James and Laura Rhodes and family (for continuing to take care of your military folks), the Eberts, Kelly and Bobby Jones, and the Artz Family.

A huge thank you to Hank and Amy for the expedient editing job. You guys are wonderful and so dear to my heart.

To my three crazy teenagers—Claudia, Joey, and Cecelia—who took care of me with all of the support, understanding and love that a mother could ask for. They also did a very good job of cleaning their rooms and putting away their dishes while I was away. Please, if you see them on the streets—tell them to get out of the street—then praise them for being (mostly) easy on their dad while mom was away.

Speaking of their dad—to my dear husband, my rock, and my love, thank you for holding down the fort and not complaining once. And for keeping me sane on many, many, many days.

To my family—parents, in-laws, and sister Liz—thank you all for listening to me on my drives every day.

To my sister Kaye (and her husband Logan) who offered to buy my whole team pizza one night at the armory. And thank you for all of the wonderful support that helped keep me sane on many days.

To the best of friends—Anna, Christy, Carolyn, Marie, and Kaye (and their families) who offered anything they could at any time, including more goodie bags of happiness. They hugged me when I needed it, despite the fact that I could have been laced with COVID and other street-rat diseases. They would drink wine with me on our porches every chance we could get, and who helped keep me sane on many days.

To my work BFFs—Tom, Mollie, Ciara, Amanda, and Walt—you all were sounding boards for me at different times and I appreciate it more than you'll ever know. It helped keep me sane on many days.

To Nick, my fabulous co-author and friend who motivated me to get this book written and helped me do it along the way. Thank you for always expecting—er, motivating me to do better. And for either driving me crazy or keeping me sane on many days.

To my amazing MSgt. I wouldn't have made it without you.

To my contributors who helped shape this book: MSgt, Mike, Louis, Big Daddy, Weather, and M.

Finally, thank you to all of the other incredible people that helped keep me sane on many days—whether they even realized it at the time or not. LT JT, Ying, Ripper, Chalk Commander, Chalk Leader, Mike, Roland, Tayte, and the General.

And a special thank you to wine and bubble baths for also keeping me sane on many days.

PROLOGUE

"I'm telling you; they don't want us there. They hate us."

"I mean, I'm sure the tensions are just really high right now."

"No... they hate us. There was this guy the other night... it was terrible. I can't even repeat it... And our guys up front—by the White House, they are getting hit with bricks and bottles of urine. It's insane!"

"Seriously? Wow. Yeah I definitely haven't heard that in any of the news reports."

"I'm beginning to think you're only hearing half of the real story. They are saying we have guns. We don't even have pepper spray to defend ourselves."

"It really makes me want to post something that tells people not to believe anything they hear—especially on social media. Question all sources, because I'm learning quickly that everyone has an agenda. I can't speak for what's going on in the rest of the country, but I can tell you what's happening and what I'm seeing on the streets of DC right now."

I paused for a moment, trying to gather my thoughts. I knew this mission was going to be physically challenging. My time in the Navy prepared me for the long, exhausting days that they would often bring. However, I was not even close to being prepared for the emotional challenges that this particular mission held.

Early May 2020

An Inequality of Sorts

It is easy like a Sunday morning out on the water. Two kayaks bobbing up and down in the river close to the bank. The sun is peeking up and the crisp air is giving way to what is certain to be a warm sunny day. The wind blows a bit harder off the bay. This causes the waves to pick up a bit. The windswept waves make the water a little choppy. Our kayaks roll up and down over the waves and we steady ourselves, easily leaning into the waves as they repeat for a couple of minutes. No reason to panic, as it is only momentary. Then, as quickly as it started, the wind dies down again. The waves continue, only without the wind, they grow smaller and smaller with each passing minute. The water grows calm. The waves die down and eventually go still. It is cyclical but predictable. Being on the water is a calming force. It is easy to breathe, easy to relax. The air is cool, quiet, and smells sweet. This is simply my happy place; where I long to be when the world around me is not as forgiving.

Up above, an osprey flies around with an air of confidence that only raptors have. They stand out amongst other birds with a commanding presence. No flock, no obnoxious call, just peaceful and powerful flight. Purposeful action and effective in the pursuit of their goal which is often the next fish they will catch. Watching the osprey is also calming, reassuring me that nature has some things figured out. The ecosystem of the river has a balance that has been struck.

Chaos and strife seem like they are strangers to the serene setting of the morning. It is impossible to imagine how different things are as I think back at this very moment. What the world is like now, compared to what it was on that calm morning just 60 miles south of the Nation's Capital.

Monday, May 25, 2020

The Waves of Injustice

It all began at the end of May in my home state of Minnesota. A seemingly routine police call on a suspected forgery in progress ended up with the death of a man named George Floyd at the hands of a few Minneapolis Police Officers. The Officers detained Floyd while someone filmed from their phone. One Officer restrained Floyd by putting his knee on the back of his neck for almost 9 minutes. Floyd went unconscious and later died as a result of this action. The video was shown across the world. In the video we all witnessed the despicable acts of the police officer as Floyd repeats that he cannot breathe. We all see George Floyd pass out. We hear from the news reports that he later is pronounced dead at the hospital.

Tuesday, May 26, 2020

The Blowing Wind

News of George Floyd's death is making its way around the world. The eyes of the nation and the world are on Minneapolis, Minnesota. Almost immediately, the calls for the arrest of the officers can be heard in the buzz that surrounds the incident. The greatest country in the world is now on tilt, some reacting to their feelings and others waiting to see what will happen next. The freedom of the people is in question. The question of racial equality is again at the forefront, as it has been since the founding of this great nation.

What makes this wind particularly forceful is that it comes from an unusual stormfront. This storm was born in what had been probably the most unique starts to any year in recent memory. The country, the world even, had been in quarantine for over two months at this point; ordered to stay home by local and national authorities because of an outbreak of a contagious and deadly virus originating from China. The popular name for this virus is COVID-19. Schools across the nation cancelled in-person classes for the year. Restaurants, bars, department stores, and small businesses all over the country were ordered to shut down to help prevent the spread of the disease. The economy took a hit, and people lost their jobs. We were a nation of people sitting and waiting for the "all clear" for an uncertain amount of time. There was a growing strain of uneasy across the country that had not been felt in a very long time. Protests for reopening the country were cropping up. Governors acting on behalf of their states made decisions on timelines and got caught in a national debate of how to handle the return to normal. Little did they know that the COVID-19 disease was the biggest story of the year—probably the biggest story in a generation. And when the story of George Floyd's death usurped the COVID-19 reels on the evening news, there was a feeling that this time the waters were not going to be calm any time soon.

At the time of George Floyd's death, the CDC had reported 100,000 deaths in the United States attributed to COVID-19 in little over a 3-month period. Later this number would be questioned as many people could have died from other causes, yet their death was attributed to COVID-19. George Floyd himself tested positive for COVID-19 in his autopsy. Incidentally, he also tested positive for significant levels of fentanyl and methamphetamine.[1]

---

1   https://www.hennepin.us/-/media/hennepinus/residents/public-safety/documents/floyd-autopsy-6-3-20.pdf

Wednesday, May 27, 2020

Sparks Fly

It is often said that one life lost is too many. Well in this case one life lost was just enough. The pressure cooker of an economic shutdown and people in quarantine burst into a scene that unfolded on the streets of Minneapolis and was broadcast across the world. The numbers of protesters who took to the streets grew, as did their anger towards the perceived unjust actions of the Minneapolis Police. Some of the protesters turned quickly into rioters who burned and looted buildings across the city. Later in the evenings, the crowds went from angry but civil, to furious and unruly. They grew increasingly violent into the early hours of each day. The protesters gave way to rioters who destroyed the small businesses in their own neighborhoods that had been closed for months. They looted and burned, all the while invoking the injustice they had witnessed just days before. It was a large wave of emotion and fury that spoke to the rest of the nation and the world.

Thursday, May 28, 2020

Fire Spreads

The slow burn of emotion was not contained to the normally peaceful Midwest. Protests and riots broke out across the country. With each day momentum grew, and the winds blew harder, fueled by the chants that became louder. Echoes of the protests played out on the 24-hour news cycle with journalists embedded in the fray. Social media and smartphones in everyone's hands showed the masses real-time what was happening. The world watched as riots in Minneapolis hit a crescendo in the capture, and subsequent torching of the city's 3rd Police Precinct building on Thursday night.

Minneapolis Mayor Jacob Frey, who initially came out seemingly in support of the protest and unrest, was now quickly changing his tune. He ordered the evacuation of the precinct, which appeared to signal an early victory for those who promoted chaos. He was quoted in the *Star Tribune*, "The symbolism of a building cannot outweigh the importance of life, of our officers or the public. Brick and mortar is not as important as life."[1] So, the building burned, and law enforcement was seen to be on the symbolic retreat.

Access to that scene was brought to everyone at light speed as tweets

---

1   https://www.startribune.com/frey-unrest-unacceptable-trump-promising-action/570830002/

and snapchats spread the fire across the world instantly. Those who needed a spark found it, and the cooped-up population had a potential energy like never before. Many sprang at the chance to get out and have their voice be heard. Riots broke out that night in many cities such as Los Angeles and New York. As I watched the news from the comfort of my home, I kept one eye on my own phone. I had been keeping in touch with family and friends that were in the area, ensuring their safety. Little did my friends and family know that they would soon be returning the favor. The protests and riots had also started in Washington, D.C.

Friday, May 29, 2020

Reaction

Friday brought the activation of 500 soldiers from the Minnesota National Guard. It was the first time the Minnesota National Guard was used for a civil disturbance in 34 years. Activated by Governor Walz, they were tasked to secure the area around the 3rd precinct police station so firefighters could put out the fire that still burned from the night before.[1]

The officer who knelt on George Floyd's neck was arrested and charged with Third-Degree Murder and Second-Degree Manslaughter. The protests and violence persisted, despite the news that the justice process had started.

With the National Guard in play, the precedent was set. Again, I found myself anxiously checking my phone, not only to ensure my family and friends in Minnesota were okay, but I was waiting for my own call to action.

> *"... When the looting starts, the shooting starts."*
> —*Part of a tweet by President Donald Trump*

Many found that particular tweet from President Trump as a threat—interpreting it as, *if you loot you will be shot.* Later, we learned that the intent of that tweet was to highlight that when looters start stealing things

---

[1] 1 https://www.startribune.com/frey-unrest-unacceptable-trump-promising-action/570830002/

from stores, especially those owned by sole proprietors, the right to defend the store is often invoked. Guns drawn to protect homes and storefronts creates further violence, and often people are injured or killed.

That same Friday morning, the D.C. Army National Guard was activated. While the Minnesota National Guard falls under state authority, the D.C. National Guard has a special chain of command. Founded by Thomas Jefferson in 1802, the D.C. National Guard is comprised of less than 3,000 soldiers and airmen. They are activated by the order of the President of the United States, often delegated to the Secretary of Defense or Secretary of the Army. They are the only unit out of the 54 Guard units from all states and territories that fall directly under the President.[2]

I am a part of the Washington, D.C. Air National Guard. When the wind blows and the fire starts, it means only one thing for those of us who have taken an oath to protect and defend the constitution of the United States.

---

2 https://dc.ng.mil/About-Us/

January 2017

"I mean... you are... *old*." She said it with such delicacy. I could tell it was not her intention to insult me, she was just stating a fact.

"Excuse me?"

"No offense or anything, but you're old."

I sat with my 38-year-old hands in my lap. All my life I had been the young one. The young parent. The young one at my job. Now suddenly the tables had flipped, and it sounded like I was nursing-home bound. "Well, as far as I know, I am still within limitations of reenlistments."

"Yes... I suppose you are." The recruiter paused for a moment and looked me up and down, and then back at the papers I had filled out. Her thick dark hair was pulled back neatly in a tight bun on the back of her head. It was so tight that not even a single hair dared to escape.

"You meet all of the criteria. But you will still have to pass all of the physical requirements and training standards."

"Yes, I'm aware. Do you have the job availability at this time, or not? If not, it's fine. We can all move on from here."

What was I thinking? The walls of the small office were plastered with posters about 'Aiming High,' and 'The Sky's No Limit.' Was this as insane as it felt? I had a great life. I had an amazing husband, wonderful kids that drove me crazy in appropriate amounts, and I had a career that delivered happiness at the end of most days. I had supportive friends in my neighborhood to drink wine with on each other's porches, as we shook our fists at the cars that drove too fast past our houses. Why would I go and make things complicated again and give up one of my weekends each month?

It was that nagging feeling that I had been having for years now.

There I sat in an Air Force recruiter's office that Friday morning. Me, a Navy veteran with a 15-year break in service. A veteran that had flown in missions over countries that most kids nowadays had never even heard of. *Bosnia? Did something happen there? Kosovo? Is that next to Iraq?* Most of my friends that had stayed in the service were now chiefs and officers or were retiring at this point. And yet here I sat, contemplating leaving my suburbia, soccer-mom comfort-zone life that we had created. Why would I choose to do that at this stage in life?

After I met my husband in the Navy, I decided to get out of the service after my enlistment was complete to devote myself to my family. From the moment I held my first baby, I knew there was no way I could deploy and be away from her for six months at a time—I could barely leave the house for an hour. But now my babies were growing older and finding their own wings. Soon, they would be finding their place among the world and deciding what spoke to them.

To be fair, I *was* old. I would have to go through the Military Entrance Processing Station (MEPS) and Technical Training again with a bunch of kids—mostly right out of high school. I would have to get re-qualified and learn an entirely different branch of the service that operated quite differently. I knew it did not make sense to anyone that I told. But the nagging feeling would not go away, no matter how hard I tried to ignore it. I was not done serving my country.

Saturday, May 30, 2020

Be Advised

I woke up to an ADVISORY in my email from the Commander of Naval District Washington:

> *Today and throughout the weekend, there are multiple protests planned in and around the D.C. metropolitan area, with the possibility of civil unrest and police countermeasures. You are advised that if you see a civil disturbance, stay away from the area to protect yourself and your families. In many cities, protests have turned violent. Military members, the civilian workforce, and their families should remain vigilant; take prudent force protection and public safety measures. The COVID virus is still in the environment. COVID-19 restrictions are still in place meant to preserve the force and prevent large gatherings. If the nation calls, for any reason, we must be ready. This message is released by Commander, Naval District Washington. Thank you.*

Washington, D.C. protests had turned violent overnight. Fires were set; violence and looting were rampant. The United States Secret Service closed Lafayette Park. It was soon becoming apparent that this was not just isolated to Minnesota anymore, nor would things settle down any time soon. It was only a matter of time before the call came. I needed to get my head straight for what the coming days would be like.

> **May 30, 2020 CMR 07-20 Secret Service Statement on Pennsylvania Avenue Demonstrations Washington D.C. –**
> *On Friday, May 29, and into early Saturday, May 30, 2020, U.S. Secret Service Uniformed Division Officers made six arrests during demonstrations in and around Lafayette Park and along Pennsylvania Avenue near the White House. Demonstrators repeatedly attempted to knock over security barriers on Pennsylvania Avenue. No individuals crossed the White House Fence and no Secret Service protectees were ever in any danger. Some of the demonstrators were violent, assaulting Secret Service Officers and Special Agents with bricks, rocks, bottles, fireworks, and other items. Multiple Secret Service Uniformed Division Officers and Special Agents sustained injuries from this violence. The Metropolitan Police Department and the U.S. Park Police were on the scene. The Secret Service respects the right to assemble, and we ask that individuals do so peacefully for the safety of all.*[1]

When most people think of the Secret Service, they think of the man or woman in a dark suit, sunglasses, and the trademark earpiece. They think of the President's bodyguards. While they certainly perform that task, they also carry out a much larger role of protecting the American people as part of the Department of Homeland Security since 2003. Originally, the Secret Service was founded as a measure against counterfeiting and forgery, the same crimes that George Floyd was attempting during his final day. The Secret Service worked for the Department of Treasury for almost 140 years before making the transition to their new chain of command. They still primarily focus on their original mission, with over 150 offices both stateside and overseas.[2]

In the press release, it was clear that these uniformed agents dealt with direct and deliberate violence. This unfortunately set a bad precedent for the days to come and would lead to the escalation of support. The normal

---

1 https://www.secretservice.gov/data/press/releases/2020/20-MAY/Secret-Service-Statement-on-Pennsylvania-Avenue-Demonstrations.pdf
2 https://www.secretservice.gov/about/overview/

complement of Secret Service in and around the White House, combined with the Metropolitan Police and Park Police were not going to cut it. They needed more support... and quickly.

Sunday, May 31, 2020

Ready, Set, Go!

It was supposed to be a relaxing Sunday with family, but we found ourselves listening to the news all day, replaying the events of the previous two nights in D.C. Fires. Bricks flying through the air. Police surrounding the White House to protect it. It was becoming more evident as the hours passed that they needed more reinforcements.

My sister had invited us up to her farmhouse to hang out and have dinner together, as we usually did on a weekend afternoon. After dinner, I walked down with her to the horse pasture to help her with the evening chores. As I scooped up a cup full of grain and poured it into the pail, my phone dinged twice.

"Be prepared, guys. Things are heating up. Be ready to report to Andrews within three hours." It was a group text from my Master Sergeant (MSgt). She was trying to give us the heads up, but nothing was official yet. Only rumors of our activation were swirling around.

I took a deep breath and shoved my phone back into my pocket. I threw three alfalfa cubes into the pail. The farm was about a half an hour away from the city, but it may as well have been on the other side of the country. The only sound that could be heard here were the chickens clucking as they scratched for worms, and the gentle whining of the horses as they communicated their excitement of their upcoming dinner.

"Are you okay?" My sister asked me with a serious look on her face.

"Yeah, I think so. I mean, guess I'm a little nervous. But I'm also excited to get in there and help out. It seems like they are really needing it right now."

"Yeah, it's really sad what's happening. What a mess."

"I know. It's crazy. I have no idea what we would even do. We're Air Guards. We are here for air emergencies," I told her and shrugged.

"I know, right?" She closed the horses into their stalls and gave them each their bucket of food. They dove in without hesitation.

"I think I'm just really tired. If we get called in now, it's going to be a long night," I said with a sigh.

"Well I am here to help however you need. You know that. You can stay here afterwards so you don't have to go all the way home. Or I can take the kids whenever you need. Just let me know how I can help."

"I will. Thank you." I smiled. And as if on cue, my phone began to ring. It was MSgt again. I took a deep breath and answered the phone.

"It's time. Pack a bag for at least one night and get up here as soon as you can. It's getting bad. They need us in the city tonight," my MSgt directed with urgency in her voice.

"Okay. I'll start getting ready. Um, where do we go?"

"Just meet at the Wing. They'll be sending out a message in just a few minutes with all of the details."

"Um, okay. I guess I'll see you in a little while." I hung up. The moment was here. I suddenly was having a hard time catching my breath—and I froze, unsure of what to do next. *Do I grab the kids or leave them here? Do I finish watering the horses? Where is my husband? What do I even pack?*

"Stop. Take a breath," my sister instructed. "Let's get the kids in the car so you can go home and pack."

"Okay, yeah. That makes sense."

And then she pulled me in and hugged me. "Everything's going to be fine. We're here for you. Just be safe, okay?" She said, her voice wavered just a bit.

I took another breath. "Okay, thank you. I love you."

"Love you too. Text me when you can," she instructed.

"I promise." I ran up the hill to grab my family, and we took off for home so I could change into my uniform and throw some toiletries in a bag. I had no idea what we were headed into, nor how long we would be gone.

Sunday, May 31, 2020

* * *

The D.C. Air National Guard was ordered to join the already established Joint Task Force. The Joint Task Force District of Columbia (JTF-DC) is normally activated when there is a planned major event where security needs to be tight and widespread. The need for coherence amongst the dozens of entities who have jurisdiction in the District is important. JTF-DC is also activated when there is an emerging situation that needs coordination amongst the groups. To look at the list, it is amazing that there can be any coherence when you have organizations that are used to doing different missions with different leadership and chains of command.

First, you have the different police organizations. The Capital Police, Park Police, and Metropolitan Police are all listed as part of the Task Force. The Secret Service and the wider Department of Homeland Security are another key part, bringing the Customs and Border Patrol into the mix. The Drug Enforcement Agency (DEA) and Defense Intelligence Agency (DIA) add to the alphabet soup of it all. Finally, the Military is involved, falling under United States Northern Command (NORTHCOM) with the Army and Air Force involved.[1]

* * *

I spent the evening and into the night at Joint Base Andrews getting medically screened for COVID-19, and waiting in an endless line for our gear from Supply (bulletproof vest, gas mask [filters], and helmet). The entire Wing had been activated for all available personnel, so the lines wrapped around the outside of the buildings. We had nothing but time to kill as we waited, so we made jokes about them sending us, the Air Guards, out on the street. We could keep the fighter jets maintained, launched, and flying in the air, protecting the city in a moment's notice—but keeping crowds from burning down the White House, this was not exactly our specialty.

I made my way into the building and finally arrived at Supply. The first question that came from the guy that were handing out the gear was what size bullet-proof vest did I wear.

"Ummm. Well..." I stuttered a bit. I can honestly say that is not something I had ever been asked before. The guy looked impatiently at me, awaiting a better response.

---

1   https://dc.ng.mil/Components/Units/Joint-Operations/Joint-Task-Force-District-of-Columbia-JTF-DC/

"*Probably* not large—if it's a men's large. Maybe medium? Do bulletproof vests run on the smaller side or larger? And would you rather it be a loose fit or tighter one?"

"Ma'am, just try this medium one on."

The supply guy held up the contraption. I stood there, yet again, looking puzzled.

It was nearing 0100 (1am). I am sure he was tired and the last thing he felt like doing was answering my bulletproof vest questions. But seriously, I could not have been the only one here that had never worn one.

He sighed and held it up and over my head. "Here, put your head in it this way." He draped it over my shoulders and let go. I stumbled and nearly lost my balance from the weight of it. I was not expecting it to be so heavy.

"Oh my gosh, how much does this thing weigh?"

"With all the inserts and attachments? Probably close to about 80 pounds."

"80 pounds? Seriously? That's just insane. How am I ever going to maneuver with this thing on?"

Ignoring my question, the supply guy began to strap me in. "Okay, well just Velcro it here and here. Pull this tight over here. And wrap your sides around here and secure them." Then in seconds, he whipped up some crazy combination of straps and suddenly I was secured. And I had no idea how I would ever repeat this on my own, let alone get the darn thing over my head on my own.

As soon as the first 200 people went through the massive line, they were sent outside to begin riot-control training. They would be the first ones out on the streets.

Somewhere around 0130 (1:30am), when the second half of us finally made it through the lines and at last had received our gear, they told us that things in the city had been covered for the night and that we were to go home, and to report at 1300 (1pm) the next day. It was both disappointing and relieving.

As I drove home that night I was lost in thought. I still did not know exactly what we were going to be doing, nor what to expect out on the streets. The uncertainty made for some wild imagination of violence and chaos. Just me, my bulletproof vest, and my helmet against a potentially hostile crowd. No weapons, just us standing there attempting to protect the peaceful citizens and preserve the law and order with wishes and magic.

I ended up arriving at home somewhere around 0230 that morning. I was relieved to be in the comfort and safety of my home, if only I could find the same peace in my mind. The unknown was my worst enemy.

May 31, 2020 CMR 08-20 Secret Service Statement on Pennsylvania Avenue Demonstrations Washington, D.C. —
On Saturday, May 30, and into early Sunday, May 31, 2020, U.S. Secret Service Uniformed Division Officers made one arrest during the demonstrations near 15th St. NW and Pennsylvania Avenue. Some demonstrators repeatedly attempted to knock over security barriers and vandalized six Secret Service vehicles. Between Friday night and Sunday morning, more than 60 Secret Service Uniformed Division Officers and Special Agents sustained multiple injuries from projectiles such as bricks, rocks, bottles, fireworks, and other items. Secret Service personnel were also directly physically assaulted as they were kicked, punched, and exposed to bodily fluids. A total of 11 injured employees were transported to a local hospital and treated for non-life-threatening injuries. No individuals crossed the White House Fence and no Secret Service protectees were ever in any danger. The Secret Service respects the right to assemble, and we ask that individuals do so peacefully for the safety of all.[2]

---

2   https://www.secretservice.gov/data/press/releases/2020/20-MAY/Secret-Service-Statement-on-Pennsylvania-Avenue-Demonstrations-May-31.pdf

March 2017

"Are you going to have to go to war, Mom?" I turned around to look at my youngest daughter's little angelic round face. Her bright blue eyes were nearly on the verge of welling up with tears.

"Oh my gosh, Cecelia. No. The chances of something like that would be so small. The National Guard's job is to take care of everything here in the U.S."

"And you're not flying anymore like you did in the Navy?"

"Nope. No flying. I'll only be gone one weekend a month—and on those days I will be home by dinner time."

"So, you won't be in danger?"

"No, hon. I promise I'll be fine. Going back into the military is something I have wanted to do for a while now. I just really miss it. And now that you guys are all older, it just seems like it's the right time."

"Do you get a big gun, Mom?" My son piped in. His thoughts were always beyond the simple, and usually involved fires and explosives of some sort, as most boys did. At least that is what I told myself when I tried to reassure my own concerns about that kid. (I never had brothers.)

I laughed. "No, Joey. I'm going to be out taking care of the flight line and runways so that the fighter jets that protect D.C. can safely take off and land.

"But I might have to miss a few soccer games or football games though. I need to make sure you guys are going to be okay with that, because if I join the military, the whole family is joining the military. That's just how it works."

"Yeah, Mom. Dad goes away for work. It's no big deal." He paused for a moment and looked out the truck window. A few seconds later he turned back towards me. "But can we come and see the jets?"

"Of course, Joey," I replied, and took a deep breath. "You can definitely come and see the jets." I smiled at the thought of it.

Wow. This was really happening. Again.

Monday, June 1, 2020

Boots on the Ground

Curfew in effect from 7 PM to 6 AM

DC NATIONAL GUARD PRESS RELEASE
WASHINGTON, D.C. – *Soldiers and Airmen from the District of Columbia National Guard have mobilized in support of the District's response to increasing protest threats across the city. The DCNG was federalized by the order of the President of the United States.*

Hundreds of Soldiers and Airmen are supporting U.S. Park Police, federal police, and the Metro Police Department to maintain order during protests in the vicinity of the White House and federal monuments. DCNG members are utilizing crowd management gear and may be armed for personal protection.

Additionally, National Guard units across the country have been immediately deployed to support the DCNG's mission in the city. All guardsmen are in Title-32, 502 federal status.

The National Guard personnel are trained, equipped, and prepared to assist law enforcement authorities with protecting lives and property of citizens within the District. This is our home, and we are dedicated to the safety and security of our fellow citizens of the District and their right to safely and peacefully protest.[1]

---

1 https://dc.ng.mil/Public-Affairs/News-Release/Article/2204415/district-of-columbia-national-guard-mobilized/

One week removed from George Floyd's death and here we were bussed to the Armory from Andrews Air Force Base. We spent the day getting issued our Riot Gear (Big body shield and helmet face shield). Now, I work out *almost* every day, lifting weights and running on the treadmill (to keep up with my taco and wine habit), but this was a completely different kind of exhausting, and it worked muscles that I did not even know that I had. We were given parachute bags to carry our gear (bulletproof vest, gas mask, and helmet). They were heavy, green canvas bags that were just a wee bit too long when someone of my height of 5'4" held it by the handles, so I could not just carry it with my arm down at my side or it would drag on the ground. It was a constant bicep workout. Before long, my arms began to burn from the weight of the gear, and I would have to switch positions. Eventually, with enough trial and error and cursing, I found I could go longer without a break by carrying the blasted bag over my shoulder. After a while I would have to switch shoulders, often losing my balance in the process with the weight of swinging the bag around. It felt like we carried these bags for hours upon hours that day, from our car to the Armory. Down the stairs to get issued more gear. Up the stairs, out to the field, back to the Armory. It went on for what felt like forever, and I soon grew to hate everyone around me while I carried the bag.

In front of the Armory were the armored vehicles parked all in a disciplined line that we had to pass through to get to the front lawn. On this sweltering afternoon, we had a quick 'Riot Control for Dummies' version of training led by an Army soldier. Thankfully, he was very patient, as he had to assist some of us with just attaching the shield onto our helmet. I did not want to be the dumb Airman that could not attach her riot shield on her helmet properly, but I specifically *could not* attach my riot shield on my helmet properly, nor even knew how to raise and lower it until the soldier helped me. Did I mention I am in Airfield Management? This was unlike anything I had ever done before.

The leg shields attached with Velcro, and I quickly discovered it was a one-size-fits-all (in theory). The top of the shield came up to the top of my leg, as the section of the guard that was supposed to cover the boot top extended an additional six inches beyond my boot, like a giant pair of clown shoes. The instructor shrugged and said that part was removable if we chose to do so. Someone behind me shouted, "I wouldn't remove that; your feet could be easily crushed." The problem was, I knew I would be doing a lot of walking. It was nearly impossible for me to walk without tripping on these things. Fall on my face every few steps or possible crushed feet? I ran the statistics and quickly threw my boot covers in in my bag.

Once our entire body was covered with a barrier of some sort, the soldier began to go through each of the commands that we could hear, and different scenarios that we could encounter. He taught us how to hold our body shields to form a wall if things are being thrown at us. He taught us what to do if someone tried to assault us from above or below the shield. And as he went on with all of the tumultuous possible scenarios, the reality of the situation began to creep in. Occasionally, I would have moments of fear in the uncertainty of it all... just for a moment. And after I took a deep breath, and allowed it to pass, I would remind myself why I was here, and all was good again. Mostly.

Finally, after all of the training, we headed back into the Armory for further instructions from our group commanders. At that point we were divided into Chalks. Yes, I said Chalks. If I'm being honest, for the longest time I could not tell if they were saying Chop, Shop, or Chalk. After I found out it was Chalk (which just sounds weird), I actually had to Google what it was, as I had never heard the term before. Apparently, it is a group of paratroopers that deploy from an aircraft. So, no. That did not really clear things up. Eventually I figured out that our Chalk was simply a smaller military group. Fortunately (or perhaps unfortunately), we did not get to jump out of airplanes into the city. Though that would have made an unforgettable entrance.

Then in a great anti-climactic wave, my Chalk was told to leave our riot gear neatly stacked on the Armory floor as we would not need it tonight. Our orders were in, and we were headed out on monument duty: guarding the monuments, as they had been vandalized (and then cleaned up by the Utah NG) over the weekend. The Park Police did not want to risk that happening again. *Monument duty* did not sound as glamorous as the front lines. I assumed the pace would be much slower, but I knew it held its own importance. There was something significant about protecting our history.

With just our IBS gear (bulletproof vest, gas mask, and helmet), we loaded up onto the bus. Full disclosure: I also had to Google IBS. Army IBS. Army IBS gear. IBS gas masks. It was a bunch of dead ends into descriptions of medical conditions I really did not want to hear about. But no descriptions of battle gear. I even asked my friend in the Army. He had heard of IBA (Interceptor Body Armor—which sounds like it would make sense), but no

IBS. So, it continued to remain a mystery, but we continued to call it IBS.

Although it was becoming increasingly daunting, we were all still attempting to wear our COVID face masks (aka germ interceptors), though the virus seemed to slip further into irrelevancy with each passing day and hour. And as we were packed into the bus, two to a seat, 'social distancing' (the buzz words of 2020) quickly grew into an impossible thing of the past.

I squished into the seat with my MSgt. She smiled frequently but was mostly quiet until she needed to be heard, so we knew when she spoke that it was important. She was extremely intelligent and had no problem standing up for those that worked for her when the situation called for it. She was a leader.

The bus ride into the city was somber. We were all sweaty, tired, and really did not know what we were going to be faced with, or how long the night ahead was really going to be. My motivation was beginning to fade, and I could sense it fading in others around me as well.

Then, over the bus radio that played on in the distant background, I could faintly make out the words to something that could only be described as a... *Cheap Trick*. Now I don't care who you are. If you grew up during the '80s, I distinctly believe it is near impossible to not sing along with a song such as this. However, I think I was about one of the five of us on that bus of 44 that actually grew up in the '80s. Good thing that I did not care.

"I want you to want me..." I started singing to my MSgt, who just smiled and rolled her eyes. She was used to this. So I got louder. "I need you to need me... I'd love you to love me... I'm begging you to beg me..."

The one advantage of having a mask on was that no one actually could see my lips moving. So really, they could not possibly know for sure that it was me singing. So, I kept getting louder, with full intention of making her laugh... and quite possibly distracting myself in the process.

"Didn't I, didn't I, didn't I see you crying? Oh, didn't I, didn't, didn't I see you crying..." She could not help but laugh—we both did. Afterall, it was nice to escape, even for a minute.

"We're going to have so much fun tonight," I told her. I mean, it was a nice thought at the time.

When the bus came to a halt, and we exited, our group split off in two, and began to encircle the Washington Monument from two sides. Walking off with all my gear into the grassy event grounds that surround the monument was not anything like what I was expecting. There were no fires. There were no crowds shouting profanities at each other with their fists up in the air. Instead, we were met with people out for an evening jog on the sidewalks. There was a couple sitting on a blanket together in the distance. People were riding bicycles up and down the paths. And in the distance, there was a young girl in full cap and gown, holding balloons in one hand, as she was in the middle of having her graduation picture taken with the monuments in the background. Where was the chaos and the fires that the media was reporting? It was not here. At least not in this moment.

A few people were taking videos of us with their phones as we surrounded the monument. It was a sight to be seen, no doubt. Every ten feet or so stood a guardsman in fatigues, in front of the flags that surround the iconic symbol of our Nation's Capital. I was proud to be there. I believed our mission was important and patriotic. And in my arrogance and naivety, I just assumed everyone else did too.

But soon, the hours of standing guard quickly lost the appeal and grew long. Eventually, as I was not as prepared as I should have been, I ran out of my bottled water while standing directly in the sun. As time lingered on, I began shifting my weight from leg to leg. I would grip my left wrist with my right hand and use my forearms to push up on my vest to relieve my shoulders. My t-shirt under my fatigues was soaked with sweat and clung to my skin. Because there was absolutely no air circulation with my vest on, it would remain wet the rest of the night. As I rested my chin on my vest, I could feel the heat rising out from inside like a freshly stoked furnace. I hated sweating, and this had to have been some of the worst sweating I had ever done... and there was nothing I could do about it.

When one is standing guard for hours, there is nothing to do but think random thoughts. 6:45. *The kids must have had dinner already. I wonder if Aaron remembered to give them any vegetables,* as he often omits in my absence. It was not a big deal, but they were all teenagers now, and I knew their vegetable intake was few and far between. *I wonder if they finished all of their homework today... Did Claudia have to work today? I hope she remembered to wash her uniform... maybe I should text them to make sure.*

It was a weird feeling to be suddenly plucked from my normal routine and with little warning placed into this new daily mission.

The sun hung low in the sky, which only made its rays more intense and directly in my face. *Damn, why didn't I bring my sunglasses?* I made a mental note to bring sunglasses tomorrow. And more water and sunscreen. When we left, I had seen a few soldiers carrying cases of water onto the bus. *Where was the bus now?*

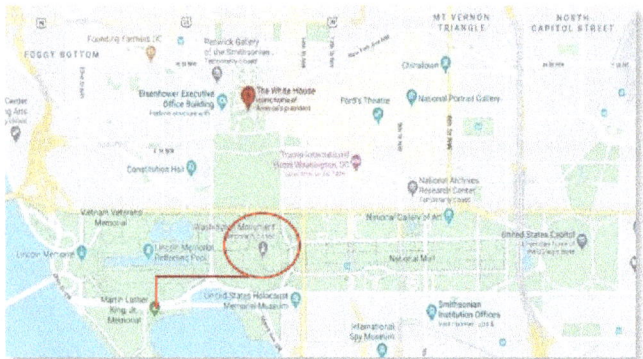

Sometime right after 1900 (7pm), our commander came to our side of the monument and pulled five of us out of our positions. "You five, go down to the MLK monument. We need coverage down there. And you're in charge," he said pointing at me, as I was the highest ranking in the group.

"Oh, wait, Sir. Um, *which way* is the MLK monument?" I smiled, hoping it was not an obvious question. The MLK monument is the park and Memorial that is dedicated to Martin Luther King Jr.

"It's somewhere on the left side between here and the Lincoln. Just look for signs—use Google maps. You'll find it. I have your cell number; I'll let you know when to come back."

"Okay Sir," I replied. Still unsure, but knowing better than to ask any more questions, I just decided we would have to figure it out. With our new orders, the five of us headed towards Lincoln Monument that rested at the other end of the reflecting pool. I was excited to move my legs again instead of just standing in one position, and to get into some shaded areas. As we went along, we passed several people out walking and riding skateboards as if it was any other night. It was conflicting, because no one was supposed to be out, as it was past the 7pm curfew, but I soon realized that some people just really did not care. And I was not prepared to be the one to tell them they needed to go home. Really, we could not do anything to them anyway unless they were causing trouble. It was a small rebellion of sorts, the beginning of people becoming untouchable by authorities.

We headed off the beaten path and broke away from the reflecting pool about halfway to the Lincoln Memorial. When we crossed the street and at last arrived, there were two gigantic stones at the entrance that we had to pass between. I cautiously walked between them and along the brick pathway into the park. As the stones opened up, they revealed the water just ahead in the distance, complete with a family of ducks swimming right along. And as I slowly turned back around, I was entranced by the momentous statue of Martin Luther King, Junior, gazing off into the distance over the Tidal Basin and into the city.

It was so serene here. It radiated a powerful existence, but also peaceful. It was as if we were in our own world all together. The five of us spread out like points of a star and assumed our positions on different corners of the park. I walked to the water and rested my arms on the railing along the perimeter, looking out into clear waters that were reflected in the dimming sky. It was growing more subdued with each passing moment, as the sun had already set. Once again, the water was my calming force, and I sat for a minute just to take it all in.

I turned back and glanced up at the face on the statue. He stood tall, 40 feet into the sky, a commanding presence. He wore an expression of wisdom that could only come from experience. It amazed me that an artist could transform stone into emotion, but this entire park captured it so eloquently.

Just then, a man walking his bicycle passed through the entrance stones and came around into MLK's view. He parked his bike, got down on one knee, and lowered his head. I realized I was staring, and looked away casually, to give him the private moment that he came to have.

Was he praying? Did he ask for advice? Or did he simply come here to take a moment to gather his strength in this peaceful place to take back to whatever he was facing in his life?

It was hard not to wonder what Martin Luther King Jr. would have thought about what was happening in our country right now, over five decades after the Civil Rights Movement.

The man stood up, nodded his head at the statue, and continued on his way.

Before long, we were joined by an infiltration of the FBI, maybe twenty or more were ordered to the park to stand guard with us. I did not mind at all, as they came with guns for reinforcements and a plethora of entertaining stories.

I did not know it at the time, but my interaction with the FBI was as a result of orders put in place by the United States top cop, otherwise known as District Attorney William Barr. He took a very aggressive approach to the civil unrest in Washington, D.C., by activating factions of the FBI, DOJ, US Marshals, ATF, and most interestingly the DEA.[2] They were all brought in to assist in the efforts in and around the District. These were not part-time professionals like our team. They were members of special units within each organization. The FBI brought members of their hostage rescue teams. The Department of Justice called up some shadowy figures that belonged to Special Operations Response Teams (SORT).

We were in the big leagues now and there was no mistaking that despite our lack of training and weapons, we were well backed up should things get out of hand.

As the guard duty around MLK was calm, we killed another two hours or so just chatting about any similarities in the places we had traveled to with our work, or the outlandish experiences we had encountered. There was a short time I thought I was going to be an FBI Agent. It was right after I had been discharged from the Navy, when I was in between having babies and still trying to figure out what I wanted to do with my life now that I was out of the service. On a whim I applied for an FBI Special Agent job and had been accepted. The admittance process continued to the point where if I had accepted, it would become an entire life altering process for our entire family. Knowing this was not conducive to being a new mom, I ended up declining the opportunity. Still, it was fun to have a commonality with these agents when we spoke about the crazy tests, interviews, or lie detector stories that were all a part of the entry process into the FBI.

Somewhere around 0200 (2am), the FBI was alerted by their command that they were able to call it a night. We, on the other hand, had a bit of miscommunication with transport services. Our bus was delayed for quite a while, and our commander directed us to stay there until otherwise relieved. So, there we sat in the wee hours of the morning, the five of us, unarmed guardsmen in downtown D.C. It was never a position I ever would have put myself in 'normal' life. But not a whole lot of this evolution was normal. *How did I end up here?*

---

2   https://www.npr.org/2020/06/01/867059312/attorney-general-steps-up-federal-law-enforcement-response-to-protests

May 2017

"Everyone entering the Air Force must be able to demonstrate the Valsalva maneuver." It was written in bold black lettering on a pink paper that hung on the wall in front of me. I sat awkwardly in a folding chair wearing nothing but a thin hospital gown, waiting to do the duck walk in my underwear. The room was stark and contained a chill that was very apparent when wearing only a hospital gown. One by one the young girls filed in—five of them, all in gowns that matched mine. As they took their chairs, I casually glanced around. They all had to be somewhere between 17 and 20. One girl had bleach-blonde hair and dark-lined eyes. She wore so much make-up that it looked like she was about to go on some stage to perform. Another girl did not wear any and sat quietly with her chestnut hair pulled back in a ponytail. I could have easily been their mother, as I had a 16-year-old daughter of my own at home.

The minutes began to drag on. All this time spent waiting caused my mind to start spinning, second-guessing everything. *What in the heck was I doing?* I knew some of my friends thought that I had to have been crazy to start this all over at my age. Perhaps they were right.

"Valsalva," the petite girl with short dark hair next to me said aloud to the room. "What does that mean?"

Before anyone else could say anything, the girl across from her quickly piped up. "It's like how you pop your ears to fly. You have to hold your nose and lean over, bearing down like you're going to have a bowel movement." And as she spoke, she began to demonstrate.

"Oh god, stop!" I waved my hands up at her. "Don't push there." I cringed at the thought. "No, no. Hold your nose and just push your tongue to the roof of your mouth and think of blowing air out through your ears. Keep the pressure in your head—*not* anywhere else in your body.

They all stopped and looked up at me, as if questioning my credentials. "I used to fly." No one said anything. "In the Navy." More silence. "Like a million years ago."

With my MSgt

Army caravan headed into the city

Guarding the Washington Monument

From our corner on Consitution

Needing to be heard

Unrecognizable Streets

Feeling the impact

Riot Gear

Photos

Before the Storm

Swearing in at the Armory

Reality Check

Doris, my Guardian

Caravan to the Capitol

In the back of the LMTV

Early Hours of June 2nd, 2020

I was hot, so incredibly hot that it was impossible to cool down. I was exhausted and thirsty. And now we had to wait for our bus—unsure of when it would actually come to get us. The weight of my gear was beginning to leave bruises, and my boots began to wear blisters on my ankles. I sat down on the concrete bench in front of me to take the weight off of my back. My friend quickly informed me of the cockroaches that kept scurrying around my head. *Ugh, Seriously? Cockroaches?* I jumped up. I was running out of steam. It was all getting to be too much. I decided to sit on the stone ground instead. This way I would hopefully be able to see the cockroaches as they came at me, and it would give me time to move.

As I sat down on the ground and looked over at the wall in front of me, I read it for the first time that night.

"Injustice anywhere is a threat to justice everywhere. We are caught in an inescapable network of mutuality, tied in a single garment of destiny. Whatever affects one directly, affects all indirectly."

The words of Dr. Martin Luther King Junior were right in front of me, as if he was speaking directly to me. The words that were spoken over fifty years ago were now so applicable in this time. Were we doomed to repeat the same history? After fifty years, were we progressing, or in fact, regressing? I did not know the answers to this. But I realized in that moment there was a reason we were here, even if I did not entirely know why just yet.

* * *

We did not get picked up by the bus until 0330 (3:30am) that morning. There had been a miscommunication with Dispatch, as there was not really one established yet to coordinate the pick-ups and drop-offs of all the troops that were spread out over the city. Eventually, we were bussed back to the Armory.

After we debriefed the events of the night and were instructed to report back at 1400 (2pm) the next day, I climbed into my truck and proceeded to drive an hour to get home. I thought of the offers that my friends and family had previously bestowed on me.

"Call me if you're tired."

"You can crash at my house, so you don't have to drive all the way home."

And even the LtCol said we could stay at the base hotel, and we would be refunded so we did not have to drive so far so late at night (or rather, so early in the morning).

But I *needed* to go home. I needed that time in my own bed, around my own family to feel normal and get reenergized. And the crazy thing was, I was not even tired for the drive home. I had been on heightened alert for so long now, that my mind needed the decompression time. I welcomed the drive. I simply cranked up my music and allowed it to dissolve my thoughts. I tried to make sense of everything that was going on. It felt like the city had kind of exploded all at once and we were thrown into the fire without much warning. Was I the only one that was struggling with this? I could not help but wonder that my fellow guardsmen were thinking as we all were attempting to navigate through our new missions.

* * *

"After a quick turn around and little sleep, I was back in the armory where we received guidance to wait for further guidance. I spent most of the morning just waiting to know what we were going to be doing next. There were several instances in which we were told to be ready to go out into the city, but nothing ever materialized. It wasn't until after lunch that we were told that we would be going to pick up military vehicles from another point in the city.

Early Hours of June 2nd, 2020

"I hadn't been in an armored vehicle since my first deployment in Qatar and I never thought I would ride in one through a major U.S. city. I happened to sit next to the commander of our shift and found out quickly that I was in charge of directions. However, I didn't realize that my GPS was still on bicycle mode and we ended up taking the most indirect path to get back to the Armory.

"Within ten minutes of driving, we found ourselves on MLK Drive in D.C., which is not where I would want to place a parade of military vehicles during one of the biggest protests in American history. However, it did make for an interesting drive with only a few people yelling at us and mostly people just filming with their phones and one random homeless man peeing on the side of the road, which gave us a good laugh. Quickly, I realized that I wasn't giving the best directions and didn't know what was going on with my GPS. It wasn't until I nearly took the parade onto a running path for me to realize that my GPS was on bike mode. I quickly fixed the issue and got us back to the Armory. After this eventful ride, it ended the day for the day shift and around 5pm, I was told to be back at 0800 (8am) the next day, which was greatly appreciated because after only two hours of sleep, I was extremely tired. I did get to see all my work friends come in for the night shift, not knowing what they were going to face in the coming night. I just knew for me that my day was over, and I was ready to see my wife and go to bed."

— M, Aircrew Flight Equipment

* * *

While we were sleeping the Utah National Guard, among other guard units, took the watch around the city. They notably took on the cleanup duty. They scrubbed graffiti off of the monuments and attempted to restore the beauty to a tarnished city. Having Guard units from other states built a coalition of Americans protecting their Nation's Capital. The Utah Guard in particular was excited to be a part of this mission. They mobilized in 5 hours and spent a week supporting the call to protect their fellow Americans during the civil unrest. A frequently deployed unit, they took on the jobs many probably would not normally choose: cleaning, scrubbing, and restoring the city as close to normal in the middle of the night, only to sometimes have to repeat the process the very next day. One step forward, two steps back, but happy to serve, nonetheless.[1]

---

[1] https://ut.ng.mil/Site-Management/News-Article-View/Article/2220420/utah-army-national-guards-monuments-men-return-to-salt-lake-city-following-dc-c/

Tuesday, June 2nd, 2020

12th & Constitution

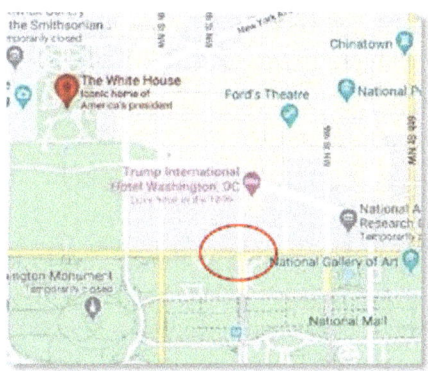

Our Chalk of 40ish people was divided into three different groups and handed different street assignments. Our group exited the bus somewhere near the west end of Constitution Ave. We began our walk eastbound, dropping off three guardsmen at each intersection that were already manned by a few policemen or another federal entity, until we were all eventually dispersed. The sweet scent of magnolia blossoms drifted down the street, as it was lined with the beautiful blooming trees this time of year.

As we walked along, there were several people that would stop and video us. It was weird. I did not know if I should smile or avoid making eye contact. (Being the awkward person that I can be, I often did both.)

One older man we passed stopped, looked at each of us walking, and one-by-one said thank you to each of us for being there and to be safe. He said, "I'm a Vietnam Veteran, and I really appreciate what you gentlemen are doing." The guys nodded as they passed him. I was towards the end of the group, and as I walked by, I smiled. He smiled back and quickly added, "and ladies."

I recognized the Smithsonian American History Museum on the right as we passed it. It was where my sister worked, and I would bring all of the kids to their once-a-month evening events to meet their guest speakers. They were innovative, inspirational people, like Kathryn Sullivan—the first American woman to walk in space. Of course, that was all in our pre-COVID life. Now all of the upcoming events had been canceled and the museum was closed to the public. It had been almost three months now, but it was beginning to feel like a lifetime ago. I never would have thought when Ms. Sullivan was autographing her biography for us, that I would be standing on this street, wearing battle gear just a few months later. When we stopped at the next intersection and our Commander asked who wanted to take this corner, I volunteered. I could still see the museum from this corner, and it felt like a piece of something familiar during this unrecognizable time.

It was a busy intersection. In front of us was where all of the traffic that filtered into the city on Interstate 395 was deposited. At the end of our street rested the pristine Trump International Hotel. Other than the White House, this lined up as number two of the most popular hot spots for a pop-up protest or demonstration. Enraged citizens stood in front of hostile crowds as their anger spewed forth onto the Trump Hotel workers. At all times a line of twenty, up to as many as one hundred, police officers would be standing as the human barrier between public anger and the cement foundation of the hotel building. Because we were doing the small favor of keeping their hotel from being burned down to ashes, the workers let us use their restrooms inside the majestic hotel. And it was marvelous.

For starters, it was simply wonderful to just get off of the street and in the air conditioning for even a ten-minute break. We were treated like royalty and extended what seemed to be genuine kindness as they thanked us for being there, and even offered us water and coffee. I had heard the rumors of this place and had to see it for myself—even though after six hours in the sweltering humidity, I had already sweated out the countless bottles of water I had consumed and really did not have anything left in me. When my combat boots stepped off of the steaming streets and into the magical world of marble floors and crystal chandeliers, it felt like I was being transformed into a mystical palace that only a small amount of people knew.

* * *

By day three of carrying this near 100 pounds of extra gear, I realized it was so much easier to wear my vest back and forth rather than carrying it in the parachute bag. This way the weight, although still heavy, was more evenly distributed, and not as painful to carry. I also realized I could remove the shoulder and crotch protectors (I'm sure that's the technical name). By pulling off these sections, I probably lightened up another five pounds or so of restraints. I also learned that rather than carrying an additional bag for water bottles and snacks—or worse, wearing a belt with an old canteen full of dysentery, I would stuff my extra supplies into my gas masks pouch, since I was required to wear that. Heaven forbid I would have to use it, the streets would have suddenly been littered with beef jerky and water bottles, as I pulled the quick-release tab to obtain my gas mask. Perhaps it is the mom in me, or the fear of not knowing when we would be able to eat next, but I squirreled away nuts, jerky, and always at least one Snickers or source of emergency chocolate into different pockets on my uniform and gear. And of course, I had to have a bottle of hand sanitizer in one of the pockets at all times, because, you know, the Corona.

Our mission on this day was to keep civilian car traffic off of these cross streets, to keep the protesters safe from getting taken out by a car. Their most assumed path went back and forth from the White House and back down to the Capitol via Pennsylvania Avenue. When we arrived, there was an LMTV (light, medium tactical vehicle), courtesy of the Army, parked sideways across the street. In my opinion, there was not anything light and medium about this gigantic military-green kick-ass-and-take-names-later truck. But coming from the aviation world, this truck—and the other Humvees that the Army sported—were like bright, shiny new toys ready to get dirty. And although I lived and breathed anything airplanes, it was always fun to play with some new toys.

On post, there were the three of us Airmen, two Army soldiers (that had arrived in the LMTV), and about four DEA guys. Drawing from my conversations with the FBI the previous night, I quickly learned that if the FBI were a gang, they would be the Sharks or the Jets in West Side Story, while the DEA were more like the Hells Angels motorcycle gang. These were the guys with the big guns, literally—some were strapped with semi-automatic rifles. These were the guys that brought down drug cartels in a single swoop. They wore regular clothes, had long hair, and unshaven beards. They were rough around the edges, and they were the ones you wanted on your side.

We all introduced ourselves and they gave us swift instructions: direct and control the traffic in this intersection as needed, and only let federal vehicles past us, or people that were staying at the hotel. And it was a funny realization, because I think most people just assume anyone can direct traffic. It seems easy enough, right?

I have learned that it is important to know what you are good at. It is equally, or even more important, to know what you are not good at. I know for a fact that I am terrible at any sport that involves throwing a ball (or a corn-hole sack). On this day, I also learned that I am terrible at directing cars. As traffic became backed up, I decided I would step in—with the best of intentions and help with the incoming cars, as the guys were busy with the other vehicles. I waved my right hand towards me. No one went right away, so I held it up to stop. *Did I need to circle my arm and point with the other?* The car on the right started to go. So I waved my left hand at the car on the left and they both went. *Oh no.* I quickly held both hands up. They stopped, but then I realized no one was going.

When directing traffic, it is very essential to be decisive. You want to make a decision and stick to it. This is not one of my strong points in life, in general. It was kind of like when you run into someone and you both go left and then you both go right, and on and on. So, I waved right to left, and then left to right. I did not want anyone to feel left out. And that did not work out so well. As this was happening, more cars were beginning to pile up behind them and soon no one was able to go. Unsure of how to untangle them, I just held up my arms, causing all of the vehicles to stop. *Well, crap.*

It was about this time that the Lieutenant walked over to interject. Very gracefully and with skilled hands, he began to untangle the mess. I did the awkward moonwalk back to the truck barrier, where it felt safe. After all was well, he looked over at me and very kindly asked, "How about you just stick to checking their IDs?" I nodded in agreement and replied, "yes, Sir."

* * *

Tuesday, June 2nd, 2020

When the afternoon sun finally faded behind the buildings, we all drew a sigh of relief as the temperature slowly began to drop. The DEA guys had kept their vehicles running and we would rotate our time in there to catch a moment of bliss in the arms of the sweet, sexy air-conditioning. This entire day/night was a game-changer from the previous one. The Army Lieutenant (LT) said we could relax our battle gear unless a group was coming through, so our muscles had moments of relief. With our clean, available restrooms, rotations of air-conditioning, and a case of water that someone brought, we were living the good life. I thought to myself, *this duty isn't so bad. If I could just catch a few more hours of sleep in the night, I could make this work.*

And then reality hit. A long-haired blonde man on a bicycle with an unshaven beard rode his bicycle past us. His t-shirt was bright and apparent, as he had written the words 'FUCK 12' across the back of it. "Fuck you!!! Fuck you all!" He yelled as he rode past us. "You all are a bunch of fucking traitors! How can you live with yourself? Bunch of fucking pigs!" He rode off down the street satisfied with himself.

I realized my mouth was open. *What in the—who talks like that—and to complete strangers, nonetheless?* I looked over at my LT. He shrugged his shoulders. "What is happening?" I asked, shook my head, and squinted at him. "And who is 12?"

"It's slang; another name for the police. Drug dealers use it as code to warn other dealers of cops in the area," the LT answered me.

"Wow. I don't even know what to say." I was dumbfounded and at a loss for words.

"Um, is this the first time you've heard them like this?" he asked me, tilting his head to the side.

"Well, yeah, I guess so."

"You should have been with us over in Chinatown last night. It was pretty bad," he said.

I knew we were not just here to direct traffic, and to make sure everyone followed the curfew laws. The thing is, one can hear the ugly stories all day long from others and from the news, but it is an all together different experience to witness firsthand. Just then, my thoughts were suddenly interrupted by garble on one of the radios.

"Alright guys, sounds like we have a group of a hundred or so headed our way." The DEA guys reported what was coming across their radios to us, as we did not have any. It was becoming more and more apparent that we were just extra bodies. We were here to be a presence.

"Get your IBS gear on," LT instructed. I donned my vest and strapped on my gas mask pack (aka snack bag) around my waist. I already had my helmet on, as I had not brought my regular cover (hat). I did not even realize it would be an option to wear, even if only part of the time. (Hence, later the permanent bruise on my forehead that developed from wearing my tight helmet for twelve hours. It took days to disappear each time I wore it; but to be fair, I do have a big head, literally.)

"We have a group of 300 or so coming our way westbound on Constitution," was reported from the DEA.

I was actually excited. And not just to break up the monotony of the night. I truly was excited to see the protest—*my* first real one here. These people were making history. I was here to support them, protect them, and ensure the good people that were trying to make a difference were safe. Someday I would be able to tell my grandchildren about this historical experience. I was proud to be a part of the change for the good.

And then they came. They were led by the police cars that were constantly trying to get ahead of the crowd to ensure the route they chose (which would alter on a dime) was free of traffic. They began to march past, chanting and waving signs with a multitude of messages. The protesters came in different ages and races. The signs they carried simply stated the words that were becoming common dialog across the nation, 'Black Lives Matter' and 'I Can't Breathe.'

There were many other powerful messages. But no message stood out more to me that day than the one that I saw a little boy carrying. He could not have been any more than 5 years-old with the sign that read, "When do I go from cute to dangerous?"

That one hit me. I looked up at his mother who was walking next to him. She, like myself and every other mother, all have no greater purpose in life than to ensure that our children are safe and grow up protected from the dangers of the world. They do not deserve to feel bad because of the color of their skin, their ethnicity, or background. It is what every parent wants for their own child. It is what we, as a country, owe to the children of our nation. No child in America should ever have to go to bed afraid or feeling bad about the way they were made. I smiled at her as she passed, hoping she understood what I was telling her in my eyes, the way mothers can communicate with each other.

As the people marched on, there were even more signs. And then, somewhere in the line of people, the thought-provoking messages and signs of determination for equality and resolve for better began to evolve into a plethora of derogatory messages towards the police, and then on to target the President. By the end, most of them just ended up reading "Fuck Trump! Fuck Trump!" The protesters walked up and down the streets chanting the words, sometimes next to their own children.

When they passed by us, they wore looks of discontent. Disgust. I tried to smile, but it was getting harder with each passing moment. I was confused, and conflicted.

And then something happened that I was not prepared for. I watched as a man came up to my LT. He had broken away from the rest of the marching protesters and stood two inches from LT's face. And he began to shout the ugliest words to him. He called him a traitor. An Uncle Tom, working for the white man. The 'N' word.

'Don't react' our commanders warned us, or it will make it worse. I am not sure how it could have been worse. I wanted to help, but I was paralyzed. What could I possibly do or say in that moment to help the situation? I did not realize I was holding my breath the entire time. I could not think clearly. But LT just stood there and took it, and eventually the man moved on.

My heart was breaking inside as I watched someone, I already knew to be a good person, be taken down, intentionally and in such a nasty way. *And for what? Why was this happening? Why were we, the military, suddenly the enemy?*

After the protesters had all passed and it was just us again, I walked over to my LT. "Are you okay?" I asked him.

"Me?" He asked as he turned around. I nodded. "Why, because of that?" He made a motion up the road in the direction that the protesters went.

"Well, yeah," I replied. "I just don't even know what to do with that. I've never seen people like that."

"I mean, there's not much you can do. Just ignore it. They aren't talking to me—they don't even *know* me. They don't know what I believe in or what I stand for," he said as he shrugged.

"I know, but how can it *not* affect you? It's horrible. I'm so sorry that they talked to you like that."

"Listen. We're here to do a job. It's not personal. You just have to be stoic." He froze, imitating a statue.

"Yeah, I'm really terrible at that. I'm probably the most awkward person you'll ever meet."

He laughed and took a drink of his Pepsi. "There are some people who are just angry at whatever is standing in front of them. You can't take it to heart. You can't react. *Stoic*." He enunciated the word again.

I took a deep breath and repeated the word. "Stoic. I will try."

* * *

As the night continued, I distracted myself and passed the time by chatting with each of the DEA guys. I could not get enough of their adventurous stories. They traveled the world, wherever the job took them. They worked undercover, risking their lives in dangerous situations, and brought the smackdown when necessary. They were part of the undisclosed group in the dark alleys, away from the awareness of the country. They were always in motion, keeping America's streets just a bit safer. And all the while, we were none the wiser, tucked into our warm beds at night.

* * *

Periodically, we would get visits at our post from other entities such as the Metro Police. These guys had seriously been through the wringer in the past week. However, they came to check in on us, making sure we were good, and most often thanking us for helping them out.

At one point, a group of young, unconcerned, kids came sauntering past us. I braced myself, expecting verbal impact. There was a young lady among a group of five or so boys and girls. She was waving a sign left and right, and as she approached us, she flashed it next to one of the Metro Police officers that was deep in conversation with one of our Army guys. It was still a bit far for me to make out the words, but when I saw them conversing, I casually strolled to that side of the street, nosily, and attempting to listen in on what was transpiring—now much more on the alert than ever before.

Then I saw something that makes me smile every time the thought comes back to me. The Metro Policeman leaned in and gave her a hug, and she, without hesitation, sincerely reciprocated. I was confused but fascinated. Then I saw the 'Free Hugs' sign that she held down to her side. Such a carefree disposition she maintained, strolling up and down amidst the chaos of the times. And while we have all probably seen the 'Free Hugs' signs, and trending TikTok videos, there was something more in that moment. It was such a simple gesture. But this girl, this young girl that went up and down the divided streets of the city that night, was exactly what we all needed in that moment. She was the human factor that reminded us that we all are the same. She had a hug for everyone, no matter their color or their voting record. She was bringing colliding worlds together with the simplest action.

* * *

"Don't shoot!"

Sometime, a little after midnight, a dark figure came riding out of the shadows on his bike, slowly swerving from side to side. I could not decide if he was drunk or just carefree. But there was something about the grin on his face that gave me an uneasy feeling as he approached. I knew right away he was not worried about anyone shooting anything.

He began riding around in a slow, swaying circle in the middle of the junction. Normally a bustling intersection right on museum row, it was now silent, as the midnight hour had approached.

As much as I tried to gauge the people that had approached us the last few days, I never could quite predict how they were going to treat us. I took every interaction in stride. I smiled, as I had done my entire life when someone strolled passed me. It would be in poor manners not to. And this time, thinking he was just trying to be funny, as we were obviously unarmed, I lifted my palms to respond to his 'don't shoot' comment, thinking maybe it would make him laugh at the absurdity of it. I could not have misjudged him more.

"Put your hands down," he began in a harsh voice. "The only time I want to see your hands in the air is when I'm going at it on top of you."

Ummm. Holy cow. Did he really just say that? I was stunned. *Don't react. Don't react.* But it did not stop there. He circled his bike around and kept ranting on, even louder now.

"Oh yeah. I bet you have a tight pussy, don't you?"

All the guys I was with immediately stopped what they were doing and locked their eyes on the intruder, watching his every move.

"Yeah, it won't be so tight after I'm done with it," he laughed. Amused with himself. "I'm going to stretch it out..."

I gasped. *Stoic. Be stoic.* LT's words lingered in my head. I was trying, but I was cringing inside, and my face was reflecting it. And yet, he just kept going on. *Please make him stop, please make him stop,* I kept thinking over and over. I was the only girl. And it was humiliating.

Out of the corner of my eye I began to see my new DEA friend to the left of me, balling up his fists. We had been hearing horrible things all day. But somehow, this had taken it to an all new level. Or perhaps it was quickly becoming the final straw of the night.

Knowing that we were not supposed to engage, the guy just kept going on. "Yeah, you'd like that wouldn't you? And then when I'm done with your pussy, I'm going to bend you over and start fucking you in the ass."

This isn't happening. It was too much. I was losing my 'stoic.' But apparently, I was not the only one.

My DEA friend had had enough, and without warning, he yelled back with such fierce intensity, that it startled us all. "Hey!"

Stunned, the bicyclist whipped around to where the roar came from. Clearly, he was not prepared for us to respond.

My friend continued, holding up both of his balled up fists, walking towards him, "You'd better get the fuck out of here right now before I put this fist and this fist through the back of your fucking face."

The bicyclist was dumbfounded. He mumbled something to himself and erratically rode off like a coward into the night. Apparently, he was not expecting a two-way conversation.

My friend turned around and looked at us all, as we stood there in the silent blackness of the night. He shrugged his shoulders.

"I'm retiring soon. What are they going to do to me?"

Instantly, my emotions came flooding to the surface as I began to process what had just happened. I managed a smile for him and said a simple, thank you. Because I do not think I had much else left in me at that point. *Seriously, how did I end up here?*

June 2017

"Fit for Duty" was stamped in red ink across the front page of my papers. My medical record was created anew online this time, since at the time I left the Navy, our records were still in an orange paper folder.

With just my husband and children in attendance, I swore in the following week and officially entered the Air National Guard. From Petty Officer to Staff Sergeant. I spent many of my first months relearning new terminology, acronyms, rank structure, and uniforms. I struggled with things like saying, "Attention on Deck," and "Captain's Call" way too much. Thankfully, I had a very patient MSgt to take me under her wing and forgive my plethora of stupid questions (there *are* such things) and comments like "that's what she said," and beyond. While I missed the traditions and cowboy ways of the Navy, everything sounded much prettier in the Air Force. Barracks were Dorms. A Galley was a Dining Facility. A Lieutenant was a Captain. Everything was different. Yet it was still the same.

* * *

"Tuesday is where things began to get interesting. As we got into work that day, we found out that we would be going into the city as a security detail at the Lincoln memorial. We were posted out to the Lincoln memorial just to keep anyone from defacing any of the monuments. I can genuinely say that as

a military member, we were there to keep watch on the monuments and that was it. I was not there to stop any protests. I actually talked to several people who were out there protesting and there weren't any issues for the majority of all the days that I was out there. I had many people come up to me and several of the members I was out with simply saying 'thank you for your service' and 'thanks for being here.' That is how the majority of the day went. We posted to different sections of the Lincoln, Vietnam, and Korean memorials—just watching and making sure nothing happened.

"It wasn't until about 6 PM where things changed, and we were notified that the parade of protesters was headed to Lincoln. When the protesters arrived, I was genuinely shocked there were so many people and I didn't know how to process it. I will say that I have never thankfully been in any kind of conflict in war, and never had been in a firefight, so I could not compare it to that. What I will say is that in this moment there was an immense amount of tension, and I think that every military member and police officer there was stiff and didn't know what to think. At no point did I ever want violence to break out, and in that moment, I was just praying things remained calm.

"I was posted at the Korean memorial and just watched the protesting from afar, as it was something that I had never experienced before. At the same time that I was staring at this protest, something on the outside started becoming clear to me. I was standing beside a guy who was playing his iPod next to me and it was some kind of Motown song that kind of fit the moment perfectly. I stopped saw the protest and heard the song and it hit me—I'm in a Quentin Tarintino movie.

"That was the night that the Mayor of DC set a curfew at 7 PM, so the protest quickly ended at the Lincoln from there. As the protest was ending, we were told to quickly return to Lincoln—at which point we started cleaning our area up and were approached by several U.S. Marshals asking if we were the only ones in the area. We told them yes, and as we got back to the shift commander, found out that there was a potential pipe bomb over at the Korean Monument where I just was. At this point I had to take a moment and sit on the steps of the Lincoln and think about my life for a minute. Not that I was in any danger—it was just a tough moment to process. We ended up staying at the monument until 10pm that night, which made for a super long day not getting home until after midnight."

— M, Aircrew Flight Equipment

Wednesday, June 3, 2020

Marley's World

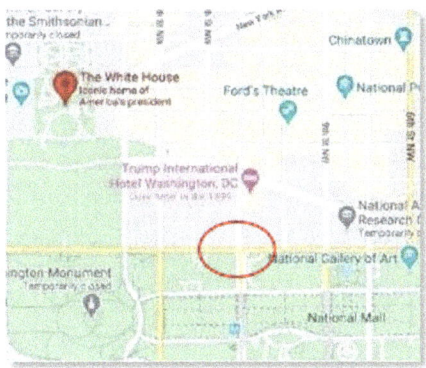

Curfew 11 PM to 6 AM

    My morning began at home with a text message from our group leader that read, "All make sure you bring your face mask for COVID or something to cover your face. Intel reports are saying when the crowds are videoing us, they are using facial recognition to find out who we are on social media." *Oh good. One more thing to worry about.*

    Our bus ride into the city was a bit quieter by day four, as everyone was beginning to run out of steam. The hours between shifts were small and the temperatures kept climbing as the week crawled on. I sat with a friendly stranger and we were both too tired to talk, so we just gazed out the window. As we drove down 14th Street, all we could see were businesses that had been forced to board up their storefront windows to prevent them from being looted and destroyed. Graffiti covered every available piece of plywood down the streets, and it was nearly impossible to find a piece that did not carry an unsettling message.

"Be happy we just want justice instead of revenge," one of them read. I felt the chills run up my spine as I read it, as it felt like a menacing promise.

Along the way, the bus came to a halt to allow people to cross in the crosswalk, but the people just kept coming. As we waited, two minutes turned into ten. There were people with signs, picketing, and marching in every direction. It was a protest marching by, but it trickled down to all corners of the street and even surrounded the bus. I just sat on the bus, gazing out the window watching them, wondering if they could see me—and if they could, what were they thinking when they saw us? I was not in the biggest hurry, as the longer I was on the bus, the longer I was in air-conditioning, and off of my feet and the blister that was growing on my ankle like an unwanted parasite. After about fifteen minutes, the protests had moved on. The street was once again open, and the bus continued along its course.

"Man, I'm not going to lie," I overheard a voice behind me tell the other airman next to him. "I was a little freaked out for a minute there. We were just sitting ducks here on the bus. They really could've done whatever they wanted to us and we would've been trapped," he said.

Like so many before it, the thought had not even occurred to me until then, but he was absolutely right.

* * *

I ended up in the same location as the previous night, performing essentially the same mission. I was with the same two airmen and the same two soldiers, which was actually quite nice. There was comfort in the familiarity. However, our DEA lineup had changed. The lead introduced himself to us as Tom and gave me some deliciously magical electrolyte powder for my water and some candy. It was like he knew food was my love language.

We also had a few new guys from the Indiana National Guard. Our federal reinforcements were beginning to grow, as more and more were being sent into D.C. National Guards all over were being activated, either to take care of the fires that were growing in their own cities or flying in to help in the

District. By my count, there were 11 state National Guards that supported the mission in D.C. Units from Florida, Idaho, Indiana, Maryland, Mississippi, Missouri, New Jersey, Ohio, South Carolina, Tennessee, and the aforementioned Utah all came with a moment's notice. Interestingly, all the augment Guard units came from Republican-led states except one. New Jersey Governor Phil Murphy took some significant criticism from local lawmakers for his decision to deploy troops, despite the New Jersey Guard being the smallest contingent in D.C. and only staying 5 days.[1] At least they showed up. Some states flat out rebuked the requests for support. Governor Ralph Northam of Virginia was quoted as saying he would not send his National Guard to the Nation's Capital for a "photo op."[2] I guess more divisive photos were not in his best interest. Pennsylvania Governor Tom Wolf simply said his Guard was "kinda busy here."[3]

* * *

Even after just two days, I was beginning to know this corner, and it was starting to feel like mine. I knew where to deposit my garbage. I knew where the bathrooms were. On our corner, we had our own broken-down blue car with the hood propped up and 'BLM' sprayed in black down the side.

This corner had its own homeless man, who was one of the most interesting men I have met. He spoke about how he loved his life, as he had the freedom to do whatever he wanted. During the day, he panhandled the corner, catching all of the Interstate 395 traffic with a sign that read 'Spare change to help a guy out?' with, 'Maybe just a smile then' on the flip side. How could people resist? Every evening around 1700 (5pm), we could smell the fragrant meat grilling on his charcoal grill as he cooked his dinner next to the street. In conversations over the two days, we learned that Marley loved his life. "I have the freedom to do whatever I want, whenever I want. Why would I change anything?" As we sat there in our fatigues in the heat, pulling Saltine packets out of our MREs (Meals Ready to Eat) because they were the only thing in that pack that I could stomach, I could not help but think that maybe Marley was onto something.

---

1   https://www.nj.com/politics/2020/06/tension-builds-between-murphy-dems-in-congress-over-governors-coronavirus-responses.html
2   https://www.fox5dc.com/news/virginia-declines-to-send-national-guard-to-dc-for-white-house-crackdown
3   https://www.radio.com/kdkaradio/articles/pa-gov-refuses-to-send-national-guard-to-washington-dc

We continued to do street check-point duty, along with monitoring the various protesters that went by, ensuring everyone was peaceful and safe from traffic. After a while, it almost felt like we were watching parades with the variety of people that passed. For one, we never knew what to expect. Most of the time they would march up and down between the White House and the Capitol on Pennsylvania Avenue, varying anywhere between 50 to as many as 2000 people. As the weekend approached, the crowds only continued to grow. There were the consolidated groups, and there were the individual groups. Everyone had something to say, whether it was expressions on their faces, on a sign, or verbally slapping us.

I was starting to think that they yelled at us with the hope that the message would be taken back to wherever it was intended. I was searching for reasons, trying to make sense of the chaos and the cruel actions that were blended in with the good intentions. At one point, a huge group all just stopped marching and sat in the middle of Constitution Avenue. The policemen on bikes lined up across our intersection to block traffic. Everyone was taking pictures. Again, there was something quite historical about it all. It was sad, beautiful, and extraordinary at the same time. And none of what was happening was anything like I would have envisioned, nor the vibe I was getting from the media.

Eventually the crowd got up, and moved on, and we resumed our "normal" traffic directing.

"So, I just got off the phone with my wife," Ying turned to me. He was one of the airmen in my unit, though I never really had the opportunity to meet him before all of this, as we worked in separate buildings. He had come over to America from China with his wife when he was twenty and gained citizenship. He enlisted in the Air National Guard shortly after, but also worked in D.C. for his 'day job.' For him, like most of us, service in the National Guard was typically one weekend a month, two weeks a year (except for the occasional deployment or activation) service.

He went on, "and she is going to bring us pizza tonight." *Hallelujah.* I had packed some turkey jerky and pistachios in my gas mask for dinner tonight, but pizza had never sounded so heavenly than it did at that moment. I had been living off grab-and-go Wawa snacks the past three days. (Wawa is our favorite gas station around these parts of Southern Maryland. And no, I have not been paid for this endorsement.)

Wednesday, June 3, 2020

"For all of us?" I asked him.

"Yes, of course," Ying answered.

"I cannot tell you how incredible that sounds right now." I smiled. Twenty minutes or so later, Ying's wife drove up to our intersection and we lovingly ushered her to the side of the road. She had brought ten boxes of pizza that we laid out in the back of the good ol' LMTV, our little private piece of shade in the street. She also bought a bag full of sodas and refused to take money for any of it. (I *am* endorsing Ying's wife right now.)

LT and I both grabbed a Pepsi like it was liquid gold. "Ahh. Now I will have good luck," he said. I tilted my head at him, trying to interpret his comment. "You know, that commercial with Kardashian girl."

I had no idea what he was talking about.

"The one where she hands the policeman the Pepsi in the middle of a protest? Like this," he reached out his Pepsi and had a whimsical expression on his face.

"No, I haven't seen it," I shook my head and smiled at the random, funny conversation I was having with my Army Lieutenant about a Kardashian girl.

"Oh man, they took it off the air or something because they got so much flack for it. But yeah, I've had a Pepsi every night now coincidentally since this all started, and I'm doing pretty good. So, it must be my good luck charm."

"Well that's a good thought. And I guess I'll have to Google the commercial."

It is weird the little things you think of in unusual circumstances. Overnight my life had completely flipped upside down. No longer was I thinking about my day-job, or baking cookies for the PTA sale (okay, okay, *buying* cookies last minute on my way to the event, running ten minutes late). I was not thinking about arguing with my teenage kids about keeping their rooms clean and to get their homework done. I was purely in a survival mode right now. I drove an hour home, slept, saw the family for about an hour, ate whatever I could grab, and drove an hour back up to the city.

I felt like I was continuously apologizing to my husband (not that I needed to, according to him, but I felt so guilty) for him having to bear the entire load of the house and family. I just did not have another minute left in my day to give. And every minute that I had with my family, I wanted it to be quality time, to make it count, although they were still living their lives, and running off in different directions.

"We're fine. Just focus on what you need to do right now, and I'll take care of the rest," he would tell me. Everything was so unpredictable and up in the air. We had no idea when it would end or what we were facing each day. It is funny, it had only been four days now, but it felt like one never-ending blur, most likely because of my lack of sleep.

But at this very moment, back on the streets of DC, we had pizza and Pepsi. There was something about it that felt really close to normal. And that was lovely.

* * *

"Is that *BACON* on your pizza?" A teenage-ish girl passed with her group of girlfriends. Her comments broke the silence of our eating.

"Umm, it's pepperoni..." I replied, slightly confused. *But bacon actually sounds better*, I thought to myself. Just then they all busted out laughing, as if they had told the funniest joke ever.

Oooh. Bacon. The cop jokes. I shook my head and smiled at how entertained they were with themselves. The ridiculous interactions were the best ones; the ones that made our night the most entertaining. But now I was just craving bacon.

* * *

"Hey, you got any more pizza?" A kid came up to our truck where we were eating in shifts, his group of three friends trailing behind him.

His question stunned me. I could not imagine taking food from a stranger on the street right now. "Um, yeah, I think we have a bit leftover." I looked up at Ying to make sure it was okay with him that I was giving away his pizza. He shrugged and nodded.

I opened one of the boxes that still had half a pizza left in it so the kid could grab himself a slice. "Oh man, thanks. Hey, can I get one for my brother over there?"

"Sure," I replied. "Why not."

He grabbed another slice. As soon as he turned to go, the other two in his group came up and asked for a piece. We passed out what was left in the box.

"Hey thanks, guys. Be safe tonight." The last one said, waved his pizza slice in the air and strolled away down the street.

"You too."

Wednesday, June 3, 2020

* * *

Suddenly Tom's handheld radio crackled and then a woman's voice came across. "All troops be advised, there is a masked man riding a motorcycle down Pennsylvania Avenue with two Samurai swords on his back." We all looked at each other.

"Did they seriously just say Samurai swords on a motorcycle," asked LT JT.

"Um, yeah," replied Tom. "Just when you think things couldn't get any stranger." He shook his head and walked into the shade. The rest of us broke into laughter. Sure, it sounded like it could be threatening. But it also sounded absurdly hilarious in that moment.

* * *

The night went on. I checked IDs. I conversed with Tom and the other guys from the DEA, still awed by another night and round of their adventurous stories. I smiled at people as they passed, but braced myself for verbal impact. I would either receive a nasty look, or a "be safe tonight" comment with an almost sympathetic smile.

After doing this same routine for a few days now, at a particular point, I came to realize something pretty significant. I found that without thinking about it, based on track record, I was beginning to stereotype. 99% of the time, I felt like I could predict the reaction of the passerby, whether they would support us or hate us. But the times that I became aware of my stereotyping, it was due to the 1% of the time that I got it wrong. The interesting part to me was that once I became aware of my prejudgment, and really thought about where it was coming from, the assumption was not because of skin color, because that reaction towards us was completely unpredictable whether they were black, white, or any other color of the rainbow. It was how the person carried themselves as they walked towards us. As most people approached us, they either had a determined, but openness about them, a kind energy, or they would come at us, already in a defensive mode, ready for a fight. There most always was an energy of sorts.

* * *

"How are you doing tonight?" I smiled as he passed. It was more of a greeting than a question. He reminded me of my best friend's dad from back home in Minnesota. He must have been in his sixties. He was tall and thin, and had narrow gray eyes. His hair was wispy and an unkept gray.

The man stopped right in front of me and stared into my eyes, making me feel all sorts of uncomfortable. "Does it make you feel good to gas people," he shouted in my face.

"Umm... what?" I was so confused and taken off guard with this one. Why did I continue to be surprised by people's actions? I did not really think about how responding to him left an open opportunity for ridicule. I think it was because there was something familiar about him. I felt like he wanted to be friendly for a split second, but then he remembered why he was here, and we were his enemy. Out of the lineup of all of us, he had stopped in front of me, the one he towered over, and went on, "I said, does it make you feel good to gas people?"

I just remained silent this time, no doubt with a confused look on my face, and most likely a bit of awkwardness. My eyes darted to the left and the right. I was not good at the 100-yard stare. Not even a little bit. I was just trying to avoid eye contact.

"You know—how *your* president gasses innocent people so he can get his photo op?"

Ooh, that's what he was talking about. The news story came back to me, which helped pull me back into the moment. Still, I had no idea how to react. What I really wanted to say was, *"um, sir, honestly when I signed up for the Guard, I thought I'd be working on the airfield and helping out with the occasional natural disaster—handing out water and food rations."*

Of course, I did not say it. I just stood there as he went on. I guess one never knows how they will react when strangers come up to them and begin blaming them for everything in the news until it actually happens. I am not a confrontational person and try to avoid conflict when possible. So here in the street, I tended to freeze up, unsure of how to react next.

Somewhere in the middle of him yelling accusations at me, LT saw what was going on and just casually took a step right in between us, even looking the other way as if he was unaware of the barrier he created. After that, the man finally decided to move on. LT looked back at me, "you know you don't have to sit there and just listen to them in your face like that, right?"

"I don't? Oh, I thought you said to be stoic."

"I mean, yeah, be stoic. But you can also just walk away in that type of situation. You're not holding any ground or anything."

"Oh. Okay." Yet another realization. Hopefully by the time I was done with

Wednesday, June 3, 2020

this all, I would know all the secrets of Nasty Crowd Control, 101. Each situation had a different vibe to it. It was strange, but in this one, I never felt like the man had a deeply rooted hatred towards us. I think he was just trying to support his cause and I happened to be his chosen target to do so. I almost had to laugh about that one. In fact, as time went on, I realized that laughing was essential for my mental survival.

Here was this guy that reminded me of a father figure, asking me if I enjoyed gassing people. I do not even send my food back at a restaurant if they get it wrong, as I do not want to make people feel bad (Minnesota roots). So to answer your question, *"No, Sir, I would not enjoy gassing anyone, ever. But I think in your heart, you already knew that."*

* * *

There is something that happens every time the sun sets on the world and the light turns into darkness. Inhibitions are lowered. People and animal personalities may alter. For some, they are more on their guard. The night is unknown, and it is intimidating. Others, though, will tend to let their guard down—as if they are able to get away with more, hidden in the darkness, like a scurrying cockroach or a rat. And my, how the rats came out on these streets at night—figuratively speaking and for real: big, hairy rats came out to dig through the garbage, and stir up anything they could find along the way. They were massive, dingy, and freaky as hell.

I was rambling on to the new group of guardsmen, getting to know where they were from and what they did normally day-to-day. As I continued to socialize, out of the corner of my eye, I caught Tom speaking quietly into his radio, his eyes locked onto something across the street. I turned to see what it was.

There were four of them, walking down the dark street, carrying themselves like they did not have a worry in the world. They were untouchable. Judging from the confidence that oozed from each of their steps, no one or thing would dare to mess with them. They could not have been much older than fifteen or so. Instantly, I thought of my 15-year old boy at home. And then I thought of these boys' mothers. *Do they know their teens are out on the streets at this late hour?*

The one in front wore a white tank top, and a baggy pair of jeans. He had black tennis shoes. I could not make out what kind they were. The others were dressed similarly, but he was the one I remember, the one that drew the most attention.

They all huddled close together for a while, talking, clasping their fists every so often. *Why did they stop where they did? What were they planning?* My mind went to the worst, and I slowly walked to the truck and grabbed my helmet to put it back on.

As I went to go do that, I heard one of our guys yell, "Watch out!" I turned to see a brick skip into the middle of the road. Tom quickly sent another update into their radio. Within the minute, an unmarked van with internal police lights hastily drove into the crosswalk of our street where we stood.

"Over there," the one guardsman pointed. "That one threw it," he pointed at the kids that were still just standing there after they had thrown the brick, as if they were waiting to see what would happen next. The van sped over towards the other side of the street where the kids were standing, and they all took off running at once. It was a real-life police chase! Well, kind of—but it was the most exciting thing we had seen all night. The kids split off as they ran, all in different directions, and they, along with the DEA van, disappeared into the darkness. More police cars flew past with their sirens blasting and lights flashing. Not sure of exactly what to do to help, I kept watch, looking to see if any of the kids would run back and try to cause any more trouble. I really did not know what to expect anymore out on the streets.

Moments later we heard across the radio that they had caught two of the guys and held them for questioning, but the other kids got away. In the excitement of it all, another minivan pulled up to the curb on our side of the street. The door opened and a few figures in uniform began walking our way. Between the darkness and the new uniforms the Air Force had recently switched to, it made reading someone's rank an impossible challenge. "How's everything going out here," the first man asked as they walked in my direction, as I was the closest of our group to their van.

"Um, good—but a bit crazy, actually. You just missed all of the excitement," I began to ramble on.

"There were these kids, and they threw a huge brick out into the road. I don't think they were aiming it at anything, but it sure seemed like they wanted to get some attention, and then the DEA guys took off after them and they all ran—and I just heard over the radio that they caught two of them." I stopped and took a breath.

"Wow! That sounds like you all have been busy," the stranger replied. It was then that I realized I knew this face and white hair from somewhere. I had seen him before. Now that he was close, I could see he was an officer, as there was an oakleaf on his cover. He was an officer and I completely missed saluting him. Then I read the name. *Oh crap.* That was not an oakleaf, it was a star, and this was the Brigadier General.

"Oh, yes, SIR." I emphasized the Sir, hoping it would make up for my lack of saluting. Then I was even more nervous and began to ramble more. "Sir, you wouldn't believe the things we have seen and heard this week. In fact, I've been writing it all down in my cellphone notes so I don't forget anything. I need to write a story about this." I wanted to stop talking, but I felt like I needed to keep going or it would be even more awkward. "I'm a writer on the side, and I have written a few books. Well, it doesn't pay the bills, so I also work for the Navy as a civilian contractor." *Why was I telling him all of this? This wasn't what he came out here to listen to.*

"Well anyways, Sir, I'm sure you're quite busy." *Stop talking, just stop talking.*

"Well that is all very interesting. Are you all doing okay? Do you have enough food and water?"

"Oh yes, Sir, we are doing very well," I replied.

"Great. Well, thank you for catching me up on everything. Please let us know if you need anything at all, okay?"

"Yes, Sir. Thank you so much." I thought about asking him for his cellphone number in case I needed anything as a joke, but thankfully, this time I just smiled and kept my mouth shut instead of making it even more awkward.

* * *

"How are you doing?" I went on with my standard greeting as a group of four girls in their late teens, maybe early twenties walked the crosswalk in front of us, their picketing signs were down at their sides at this time. It was getting late. I could not understand why they were on the streets at this time. Two of the girls had *Black Lives Matter* across their t-shirts.

"Not good!" One of the girls cocked her head to the side and rolled her eyes as she spoke.

Though I was pretty sure where she was going with this, I replied, "I hope it gets better."

She took a few steps and then turned back around, stopped, looked directly into my eyes and asked, "when this all goes down in history, when you tell the stories, who are YOU going to tell your grandkids that *you* fought for?"

In that moment the hatred that I had been witnessing towards us in this city was finally clear to me. They saw us, the military, as extensions of the president. And though we did ultimately receive our orders from him (as well as every president before him) as the Commander-in-Chief, I cannot think of a single service member that ever enlisted to fight for the President. Every person that I knew raised their right hand and took an oath to defend our entire country. Not the white citizens. Not the black citizens. Not the Hispanic, Asian, or Indigenous citizens. Every American citizen.

Don't do it. Don't respond. But I *had* to respond. I was fired up now, and without wasting another second, I looked at her as if the answer was so blatantly obvious, "I will tell them that I fought for America."

She stared at me for another second or so, narrowing her eyes, and then whipped her head around and walked away. I really do not think she could have said anything. And for the first time since I was activated, I finally felt some type of justification, even if it was only for a fleeting moment.

* * *

"Well guys, we're going to head out... are you going to be okay?"

It was 0100 (1am), downtown D.C. The weather had finally cooled down to a comfortable temperature. The streets smelled of sewer, but were peaceful, except for the rats and cockroaches scurrying about. The rats were in hog heaven, not even trying to be discrete anymore as they rifled through the garbage for their nighttime feast.

We were going on an eleven-hour shift now, and the bus that was supposed to pick us up was nowhere to be seen.

Just then my phone buzzed in my pocket as I received a text message from Master Sergeant M, our Chalk leader:

> 1:01 am MSgt M: A bus will arrive at 12 & Constitution, and when it does we will be released. So upon arrival please notify me and I can pass the word.
>
> 1:27 am MSgt M: Okay everyone head to 12th and wait for the bus.
>
> 1:28am A1C: Copy
>
> 1:50 am SrA: No bus still but more of the troops have arrived.
>
> 1:53 am MSgt M: It may be a while, some buses are diverted supporting more arriving Guardsmen. We're sitting at 7th for now.

Wednesday, June 3, 2020

1:57 am SrA: Copy

I took a deep breath. With each passing minute, my frustration continued to rise. Time dragged on as we waited. This was when it really became a challenge to remain in good spirits—the point where lack of sleep turns into deliriousness.

There were no benches around, just large, square cement planters on the street corners. A few of the guys sat on them, but I felt like I could not even hold myself up anymore. Instead, I sat down and leaned against a planter on the rock-cemented sidewalk, with my vest as a cushion. The sidewalk held warmth, even though the sun had not glared down on it in hours. My shirt was still wet from sweating under my vest, and for the first time I welcomed the warmth, as I was beginning to get a chill from the dampness.

I knew there were cockroaches scurrying around. I knew there were rats across the street. (I convinced myself that they only stayed on that side.) But none of that mattered to me anymore, because at this point, I had a total of seven and a half hours of sleep collectively since I had been activated, three days ago.

It felt so good just to take the weight off of my feet. But with each passing second, I just kept calculating how little sleep I would be getting in the night by the time I got home. I could only hope that they would give us the 'break' that they had been promising us. More troops were coming in, so we could probably start some type of shift work, and only work eight-hour days. At least that was the idea that they mentioned. *This too shall pass; this too shall pass*. I mean, it could not get any worse, right?

2:32 am Me: Did they forget about us?

2:40 am MSgt M: No, but the feeling's the same.

2:41 am MSgt M: The Commander appreciates everyone's patience. Army Trans is working it but we're not in direct contact so information is just as slow.

2:48 am MSgt M: The bus just picked up Chalk 9, And it's on its way.

* * *

When at last our bus came, I felt like I had just won the lottery. I had been conditioned to the disappointment of all the other buses hissing at me as they passed on by. At last, my street-loitering was over... for this night, anyway. Tomorrow was always a new day.

By the time we made it back to the Armory and had our debrief with our Commander, I was running on fumes. I was beaten down mentally and physically. I was near the end of my rope, and it was now approaching 0400 (4am).

"Guys, you've been great," Ripper, the Commander in charge began. "I know it hasn't been easy. I know you've been given a huge shit sandwich to sink your teeth into every night," he paused. Some version of the shit sandwich analogy was always his favorite.

"The good news is, hopefully soon, we can start making it a smaller shit sandwich. But guys, I still need you to report back to the Armory at 1400 (2pm) tomorrow."

I quickly did the math and realized I would be getting a total of about six more hours of sleep. I needed more. I could not function like this anymore. I turned around from the group, I could not help it. I was overwhelmed and a big ball of emotions. I needed to go home now. I told my MSgt I had to go, doing my best to not make eye contact. Ripper finished up and dismissed us for the night—er, morning.

"Are you okay?" MSgt asked me, already knowing the answer.

"No, but I will be tomorrow. I just have to go," I replied to her, throwing all of my gear over my shoulder and began walking to my truck.

"Text me when you get home, please, so I know you're safe," she yelled to me as we split off in different directions of the parking lot.

"I will," I promised her, walking away, as tears began rolling down my cheeks.

* * *

"The first few days I didn't even know why we were there. I actually found it a bit terrifying when nothing was really explained and all of a sudden, we were being issued our shields and riot gear in an environment that is supposed to be our hometown and safe. After we had our gear, we went into the field at the armory, where an Army soldier demonstrated how to put on the gear and initiated the riot training. And then, after about fifteen minutes the Army guy was like, "okay you're ready now." We watched all of these Army vehicles pull out so packed with military people and head into the city. We had no idea what to expect. I'd be lying if I said that I wasn't nervous.

"Our first night there we were posted at the Washington Monument. When

we filtered in and surrounded the monument, people began looking at us—almost with disappointment. As the night went on, a whole group of protesters came up across the lawn towards us from the reflecting pool and then they just stopped. At that point, more and more US agencies began to show up on the other side. There were three rows, probably 50 to 60 US Marshalls and Border Patrols. Later the Secretary of the Air Force and the Secretary of the Army both came by the monument. There were just so many different agencies. We were all wondering, what are we all here for? I understand we were showing force, but it was a bit extreme.

"There were moments over the next couple of nights that were just crazy. Flash gas canisters were going off. Helicopters kept flying over and circling around. Sirens could always be heard. I just remember looking up, thinking that this felt like a war zone.

"A few nights later, we were posted on 18th and C street. Towards the end of the night, there was a young guy on a bicycle—probably in his mid-twenties. He was riding around in front of us and would stop right in front of the DEA guys, looked at them and began shouting at them, "Somebody should kill you. **Your** life doesn't matter, motherfucker!" It was a scary situation.

"The whole concept of COVID had gone out the window. We were constantly around people—in the streets and in our own units. A lot of people asked if we even knew why we were there. I just didn't respond because I wasn't even sure some days. Ultimately, we were there to allow people to protest in a safe environment so nobody could cause harm to them.

"I'm all for people protesting. But I felt angry at the vandalism—what was the point of all of that? All of this just showed how divided we were as a country. I'm afraid for this generation. They do not have a true leader that can lead and inspire the positive changes that this country needs. I think of my daughter growing up in all of this, and it's not very comforting."

— MSgt

Thursday, June 4, 2020

Riot Training at Ellipse Park

No Curfew

FBI Director Wray confirmed that an investigation was opened into George Floyd's death almost immediately after the incident. He highlighted the apparent failure of the officers in Minneapolis to protect and serve. He reaffirmed the mission of his organization and the sacred right of peaceful protest. His most poignant statement was:

> "The FBI's mission is to protect the American people and uphold the Constitution. That mission is both dual and simultaneous—it is not contradictory."[1]

---

1   https://www.fbi.gov/news/pressrel/press-releases/fbi-director-christopher-wrays-remarks-at-press-conference-regarding-civil-unrest-in-wake-of-george-floyds-death

A great deal of official press releases surrounded the early days of the unrest. It was almost like every agency was compelled to say what they stood for, why they existed, and by what means they were involved in the movement. Almost all had a common theme: we understand why you are protesting, we acknowledge the need for social justice reform, but please do not set fires and create chaos to get your point across. But Director Wray's remarks captured why we were all there—protesters, law enforcement, and even the National Guard. It was a duality that existed because of the great nation we live in, not in spite of it. It made me proud. I was more aware now than ever that justice does not come in the form of flying bricks or flaming store fronts. Yes, it was because of that violence and rioting we were there. But we were not there to stop the protests or silence the message. We were there to protect the people and the city from themselves and itself. That was lost on many when they saw our riot gear.

* * *

The sign he held above his head in his right hand read "FREE MILK." With his other hand, he was waving the passing cars into the abandoned parking area where he stood. I was driving through South East D.C.—the route I had to take every day to get to the Armory from Southern Maryland. Along my drive I would see shops with bars on the windows, and old homes that were unkept, often falling in on themselves. The yards were full of various collections of clutter. I was advised by our Squadron's Intelligence Department not to wear my uniform on my drive in, and then advised by a friend, that drove up to the Pentagon every day for work, not to stop anywhere after dark. "Put it this way, if it's late at night, even if the light is red, you best just keep going," he told me.

But then I saw this man, and my hope was restored. In this poor area on the outskirts of D.C., there was a man with coolers full of milk surrounding him, giving it away for free, and helping those less fortunate. I thought about stopping to tell him what a great thing it was that he was doing and maybe even make a donation, but I was already running behind. But I made a mental note to stop tomorrow if he was still there. It was so refreshing to see a good-news story.

When I reached the Armory, I unloaded my gear from the truck and into my arms, and began thinking cold thoughts like ice skating and snowmobiling back in Minnesota—anything in an attempt to will myself not to sweat as I walked across the sweltering parking lot into the Armory. Unfortunately, it was a lost cause and my shirt already began to get damp.

As I walked through the doors and metal detector and into the grand opening of the central floor, I hardly recognized the place. It was in full operational swing, with troops from all over the country. Hundreds of sets of riot gear were strewn about the old wooden floor, but in organized, individual piles for a quick grab-and-go rotational use for the shifts. The bleachers were pulled out, and troops were waiting, eating, getting briefed, or sworn in for duty. There were also various tables set up around the inside perimeter. Placed across the main floor were three gigantic industrial fans—attempting, but failing miserably, to distribute cool air around the longstanding building. Despite the Armory having air-conditioning, it did not stand a chance at cooling down the main open area with the various open vents in the walls, and the opened garage door that was large enough for a semi-truck to fit through.

After I walked through the temperature screening table (because apparently COVID was still a thing) and scanned in my military ID at the next table, I turned to my right, there was now a table set up with snacks and drinks for anyone to grab—for free! We finally had a drink supply! But not just water—there were also Powerades, Diet Cokes, and coffee. On the table next to that, there were Girl Scout cookies, Chap Sticks, and Bibles. What more does one need to get through the nights in the District?

Despite the fact I had packed all of my snacks as instructed, the tables that were along the back wall contained 'to go' boxes of spaghetti from a local restaurant. Straight ahead there were two other tables that had been just set up. Word on the street was they had ibuprofen. And since I had forgotten to grab some in the chaos of getting out the door in a timely manner and back up to the city, I decided to swing by the table in hopes of numbing the blister on my ankle.

There were a few people around the medical table conversing with the nurse, so I waited a few steps back on the other end where multiple pamphlets had been fanned out. They carried encouraging messages on how to get help if one was feeling depressed. They had set up an entire table for our mental health.

I read the title on one of the pamphlets out loud. "Are You Feeling Like You Just Don't Want to Go On Anymore?"

"Um yes. Especially with this big blister on my ankle," I answered the pamphlet. I did not really think about how it sounded as I said it, until the guy in front of me quickly turned around with the greatest look of concern on his face.

"Oh, no," I quickly corrected myself. "I'm sorry, I didn't mean that the way it sounded. I'm just tired. And I don't feel like going back on the streets today. And my ankle feels like an evil troll is stabbing it with a pitchfork repeatedly."

He raised his eyebrows and slowly turned back around. *Well that ought to have convinced him that I'm fine.*

Honestly, after the craziness of the past few days—I had a hard time taking it very seriously. *Really, Air Force? Pamphlets and Girl Scout Cookies?* I mean, the cookies did help a little, (cookies always help) and I truly was appreciative of all that they were doing now, but I still could not get over the transformation from where we began just four days ago. I really do not think they saw any of this playing out the way that it had.

"Can I help you, Sergeant?" asked the nurse.

"Yes please. I was wondering if I could get some of that fabulous ibuprofen that I heard you guys had."

"Ooh," she replied with an apologetic look on her face. "Actually, we just discovered that all of our medications expired in 2017. So, I can't technically give you any of these and I just have to throw everything out. I'm sorry."

I had to laugh. "2017? Seriously?"

She nodded her head. "I'm afraid so."

"Okay. No problem," I replied and went back to prep my gear to go back out on the street, hobbling, as the evil troll continued to stab me with each step.

* * *

"Today you guys are going to be in full riot gear, not just your IBS," Ripper began his daily brief to us. "You're going to the hotspot, Lafayette Square, so you will need your helmet shields, leg shields, and your big handheld shield."

Ooh, we were finally going into the dangerous area. I was excited—but also a bit nervous of the unknown.

"Make sure you have plenty of water on you," he continued. "We will try to get around and make sure everyone is stocked with water, but just in case, take your own. Remember your crowd control training. Look out for each other and be on the alert always for objects being thrown at you. You never know what you may encounter—bricks, water bottles, bottles full of bleach or urine. And we've just been told that they are now throwing soured milk on the federal agents and military. So, keep your head up!"

Sour milk. So that's what the guy on the side of the road was doing. He was not helping the homeless, he was providing milk for everyone to throw on us all. Well I am *definitely* not helping him out tomorrow.

We turned around to walk away. "Oh, and one more thing," Commander called out to us. "Don't wear your uniforms to and from the Armory anymore. Or if you do, don't stop anywhere. There was an assault incident at a grocery store yesterday, with someone getting attacked because they were in uniform."

And with that we were dismissed to grab our gear and file onto the bus.

Sometimes it was hard not to just sit back and take in the absurdity of the entire situation. *Were we really at war with our own American people? How did we get here? Was it really us against them? How did our country become so divided? And when it came down to it, were we really that divided? At the end of the day, didn't we all just want the same things for ourselves and our families? And who the heck's idea was it to throw sour milk at us?*

I took a last bathroom break, as I had no idea when that opportunity would arise again and headed out to the bus with all of my gear. I was able to sit with my MSgt, and we both stared out the window as we drove into the city once again.

"Is it weird that I'd rather get hit with a brick than a bottle of piss?"

"Definitely not weird."

* * *

The bus dropped us off in Ellipse Park, which is the park that is directly south of the White House—the one with the National Christmas Tree. Upon departure from the bus, we were slapped in the face by another day so hot that the temperatures of hell created by the devil himself had nothing compared to this week in the District. It was well into the 90s with triple-digit heat indexes, but still we clung to the promise of rain by evening for a sweet breath of relief.

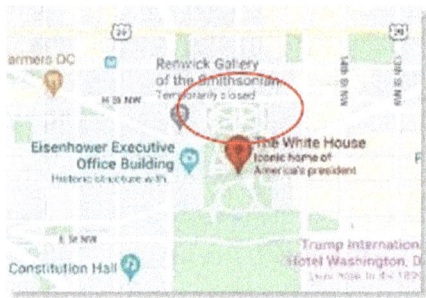

Our group, along with several others, was to be on the second shift to go to Lafayette Square Park. This was where the ongoing protesting-turn-rioting, and back to protesting was taking place. It was on the north side of the White House. Lafayette Square Park and extended two blocks or so north on 16th Street.

With the Washington Monument as a scenic backdrop, we were instructed to line up in the field to practice our street signals and crowd control training following the commands from the Army. This is where our Air Force skills began to show, or perhaps lack of marching skills. As I listened to the commands, I thought back to our December drills in which we always had our annual awards ceremony together as Army and Air National Guard. I recalled how we could always tell who was in the Army and who was in the Air Force just by their marching. I am not sure if the Army soldiers march every where they go on a regular basis, but I can sure tell you that the Air Force and Navy does not march after boot camp or technical training schools unless it is for a special event. Every year it takes us hours of practice to get back into the groove before the ceremony commences. And every year we start off with rows that look like snakes slithering, while the Army looks like rows of unwavering boards.

Out on the park lawn, there were several rows of people in uniforms and riot gear, as we were integrated with other federal entities. As I lined up in one of the back lines, I could not help but enjoy overhearing the conversation of the two Air Guards in front of me listening to the Army Guardsman give us orders.

The Army Guard yelled in an attempt to be heard by the one hundred or so troops spread across the field. "You're going to hear Air Force Split, or you will hear Air Force Stop. When you hear that, you're going to stop."

"Wait, do we Split or do we Stop?" the Air Guardsmen in front of me asked the guy next to him.

"Split means stop," his friend replied.

"Wait, what? How do we know when to split?"

"When they say Split," the friend replied.

"But doesn't that mean stop?"

"Yes!"

"Oh my god."

"Wait, what do we do if we hear 'run, motherfuckers!' Do we split or do we stop?"

And with that, all of us in the back row lost it and could not control our laughter.

Field training continued with more in-depth commands and scenarios than the previous crash-course we did on the first day that we received our riot gear. The shield that we held up was about 4 feet tall by 2 ½ feet wide and about 10 pounds. It was supposed to be a clear plastic to see through it, however it was so full of scratches and smudges that to view anything through it would appear quite blurry. To hold it, I would slide my left forearm into a strap half way up, and then grasp the handle, so I could hold it all with one arm. I was to make a fist with my right hand and place it on the shield about face height to help reinforce any impact on it. Needless to say, holding this thing for more than five minutes or so caused my forearms to start burning and shaking. Again, I was using muscles in a way that they had never been used before. I would be lying if I said that part did not make me nervous. Would I be able to hold my shield up as long as necessary if the situation grew dire? Perhaps adrenaline would kick in and I would not even notice. One never knows how they are truly going to react to a situation until they are there. And that is part of the redundancy of any military safety training. If one does something over and over, like my days back in the Navy when we would practice our 'bailout drills' (jumping out with a parachute in an emergency) repetitively, we could eventually do them in our sleep. That was the entire goal. They *wanted* you to be able to do them in your sleep in case you were in a foggy haze or a crazed emergency. It was the intention that it would just be second nature, muscle memory.

As training went on in the wide-open park, we watched as the sky started to darken in the west. Within minutes, ominous clouds began to roll in, turning the scorching sunny afternoon to black. Lightning lit up the sky, followed by the explosion of the thunder, and the downpour commenced. The commanders instructed us all to seek shelter in and around the Ellipse Visitor Center. Now holding our riot shields over our heads, they doubled as fabulous rain shields for us as we scurried to the Visitor Center.

In the parking lot, one of the supply guys had parked his truck and began handing out uniform-camouflage raincoats. Better late than never. I tucked one into my gas mask bag for when I had to go out on the street later that evening. My gas mask bag seemed to grow with more items every day. It was quickly becoming my gas mask/miscellaneous bag. I was beginning to think I really just ought to start carrying around a backpack as many of the other guardsmen did. Well, the practical ones, anyway.

* * *

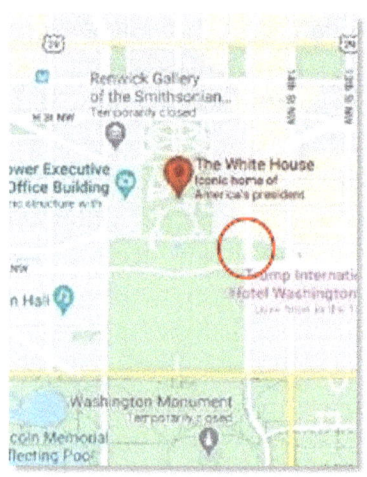

The visitor center had nice coverage from the rain all around the outside walkway and the center of the building was open on either end to the outside. The place was currently closed to the public for ongoing construction. There were two handicap restrooms upstairs that smelled like people had done very bad things in them. Thankfully, we discovered that there were also restrooms downstairs. However, it was quite a surprise when we opened the door and discovered that due to the ongoing construction, this bathroom was currently set up prison-style. There were no stall walls or doors. Instead, it consisted of toilets lined up all along the wall in a row. Now there are very open, social people that this would not bother them in the least, however, we in the Air Force are a little more conservative. It was easy enough though to just have one of the other females keep an eye on the door as we took turns using the facilities, as there only were a handful of women. *And*, I would like to add that we all had very trained bladders at this point.

While most of my Chalk sat outside around the pavilion part of the Center chatting, I went into the open breezeway to explore, as there was a lot of military gear strewn about. I was trying to figure out where it came from. I also noticed a gang of guys that were twice my size with muscles bulging out from under their black t-shirt sleeves. They seemed to command the area around them. They wore black vests that were lined with all kinds of exciting things that looked like they could cause trouble. These guys definitely were not the FBI. SORT was written across the front of their vests. *Oooh, these are guys I've been hearing about all week. The crazy ones!* I was intrigued and struck up a conversation, because really, there was not much else to do at that point while we waited for the rain to stop, or for a call for reinforcements—whichever came first.

There was a trunk next to military equipment where one of the guys was sitting. "Oh hey, do you mind if I sit here?" I asked casually as if we were at a café, or a library, or something normal, instead of next to a trunk of secret military devices.

"Not at all," he responded. He had dark skin and hair, was a bit smaller than most of the other guys. He also had the kindest smile and a welcoming demeanor, despite all of his secret weapons that I imagined he had hidden all over himself.

I put down my riot shield, took off my helmet (which felt amazing to do, as the sucker continued to leave that bruise that was now forming an indent on my forehead), and I sat down next to him. As the night went on and the downpour outside continued, we chatted on. Before long, his other co-worker and friend came by and I picked their brain with a million questions. I learned that the SORT team is part of the specialized forces from the Federal Bureau of Prisons (BOP). They had been brought in from all over the country and had an air of mystery about them. Most wore a patch from their unit and one that simply said DOJ. The teams that I met this particular night were from Georgia, and Texas. They were teachers, psychologists, accountants, and held other various 'normal' jobs. Oh, and they also happen to break up prison riots as necessary. These teams are tactically trained and specialize in disturbance control and providing assistance to other agencies during these types of national emergencies. I had to drag their stories out of them, but when they told them, I was not disappointed hearing about their adventures. So if I have not yet elaborated enough on it, one of my favorite 'good things' that continued to evolve from this time was the ongoing opportunity that I had every day to meet and get to know a multitude of America's finest from all over the country. And for that, I am grateful.

* * *

The raging storm outside continued, creating small rivers along the streets, washing away the debris and dirt in its path. We continued to wait for our call to go into the city, but the rain also seemed to wash away the rioting and unrest in the square, and the crowds dispersed for the night. The temperatures had cooled, and the air had cleared.

Someone had pizzas delivered and passed around. Another guardsman passed out the sweetest goodie bags to everyone. They were filled with fun things that make people smile such as jolly ranchers, gum, granola bars, trail mixes, and mints. Apparently, the guardsman's daughters had put these together. As it was pushing 2300 (11pm) by now and we were wet and tired, these were like little Ziplock baggies of happiness. I cannot express how appreciative we were of such a simple, thoughtful gesture.

Suddenly, an ear-popping crash shook the ground we were standing on, and a bolt of lightning shot down from the clouds and plummeted into the earth near us. We all jumped, as it was way too close for comfort, and it was obvious that it hit something nearby. Later we found out that two South Carolina Guardsmen that were in the square had been struck by the lightning! They were quickly taken away to the hospital and were in critical care, but thankfully survived. No doubt they went home with a story to beat all other stories.

It was sometime shortly after the lightning strike that it was decided we would not be going into the city tonight, as the streets at this point had become clear, and all was calm. While this was disappointing, as our day consisted of much 'sit around and wait,' by this hour, I was grateful to call it a night. Well, I should say once the bus arrived, I would call it a night. And track record called that a regular crapshoot.

Sometime after 0100 (1am), the bus arrived. As we were walking to our bus, my Chalk Commander was walking next to me.

"How are you doing, Sir?" I asked him in a sing-song voice as we crossed the street. He was also a very stoic guy. I made it my secret goal to make him laugh one of these times. But most of the time he did not get my sarcastic jokes, leaving it an even more awkward conversation.

"Oh, I'm tired, but fine," he replied. His eyes were bloodshot. I could tell he was very tired. I realized that I had been internally (and sometimes externally) complaining of our situation, and I would forget he was getting it from both sides (as leaders often did). Many of our officers that were leading different Chalks were receiving the shit sandwich orders from someone above them who were getting orders passed down from someone above them, and so on. There were way too many captains on this ship, between the Army and Air Force working in "cooperation," and everybody in the military knows shit rolls downhill. And then, as a Chalk Commander working in the field—er, streets directly with the troops, he also got to take a big bite of that shit sandwich for himself.

He looked back at the visitor center and there was still a large group of Guardsmen sitting around it. "But I can't complain. I feel bad for those guys," he gestured towards the troops. "They just got kicked out of their hotel and have to find a new one tonight."

I was stunned. "What? Why?"

"Because the hotel doesn't want the bad press of putting up the National Guard when the Mayor wants us all out of here."

I was beginning to see what a political war this all had come down to by this point. *How much of this was intentionally instigated? What was really going on? Why wouldn't the Mayor want us here to help, to protect the city and preserve the monuments from destruction?*

It almost seemed like it was a standoff between the President and the Mayor. We were just caught in the crossfire.

It was later reported that this happened because of a financial issue. According to 'fact checkers' at *Stateman News*, Mayor Bowser rescinded on Twitter, acknowledging the Utah National Guard troops were in D.C. hotels, but that "D.C. residents cannot pay" the troops' hotel bills. And blah blah blah politics and technicalities. Whatever it came down to, in the end, it was sometime around 0100 (1am) when we, the D.C. Air Guard, were dismissed to go home, meanwhile 200 Utah Guardsmen were still waiting for a place to put their bags down and get some sleep before they had to get up and work in the streets the next day. That is a fact. I personally have no idea how late it ended up being when they finally were able to get to a hotel and go to bed that night, nor do I know exactly how long they had been awake at that point. And that is quite the shit sandwich.

> *"If media reports are accurate, the decision by the Mayor of DC to cancel hotel accommodations for members of the Utah National Guard–who were deployed to restore order—is beyond outrageous. I will do everything possible to push back against this outrage against the men and women of our National Guard. They left their homes and businesses in Utah to protect homes and businesses in our Nation's Capital. The Mayor of DC should thank them–not evict them! Federal taxpayer dollars come from all 50 states to help DC. Why should we continue to provide federal funding to an entity who refuses to allow lawfully assembled National Guardsmen a place to sleep?"*
>
> —Senator Lindsey Graham of South Carolina, posted on June 5, 2020 at 10:17 on his public Facebook page.

In the absurd hours of the morning that no one should have to witness, I slowly crept through my door, unlaced my boots, and peeled them off. It was a heavenly relief. I poured a glass of wine and quietly I tiptoed up the stairs, as not to disturb the dogs or kids and slowly opened my bedroom door. As I attempted to creep across the bedroom floor, I ran into the ironing board in the darkness, causing a colossal crash. Thankfully, I saved my wine from spilling. (I'm pretty much a professional at this point.)

Startled, my husband rolled over to face me, squinting to see what was happening. "Sorry, it's just me," I whispered.

"What time is it?"

"It's 3:30am," I replied

"Oh, you're home early." He sounded slightly surprised.

"Ugh. I guess we've set that bar pretty low," I replied, exhausted. I walked through the bedroom into the bathroom. I peeled off my clothes and sat down in the long-awaited bubble bath with my glass of wine and did my best to clear my head.

* * *

"Somewhere in the middle of all of the Domestic Ops craziness, I decided to take a walk in the city one night. I just needed to decompress. I needed to see the city as a regular person, not in uniform. As I walked down my streets at

Thursday, June 4, 2020

midnight, I found it was a place I barely even recognized.

"The streets were full of litter and graffiti. And right in front of the White House was a tall black wall, keeping everyone out. It was the first time I had ever seen anything like that there. The White House used to be open for tours, year-round. But not anymore. Not much about this time was normal, and I could hardly recognize my own city.

"I cannot really explain how I was feeling during this time, other than I was very numb. It all just happened so fast, I never really had time to really process it. We had a three-hour recall and suddenly we were up at the Armory going through Riot Control training. This isn't something we have ever been trained to do.

"I didn't personally have anything thrown at me, but I had conversations with people on the street. How come you're out here, they would ask. I would just tell them that we were out here trying to protect the assets of the city—and to stop people from looting and spray painting.

"One of the days, the Army did not have enough Humvee drivers and asked if any of us Air Force had driven a Humvee. I volunteered to help, as I had driven one before. So that night, as I'm driving in the Humvee as part of the caravan, I just couldn't get over the feeling, like what am I doing? Here I was—it was amazing, actually—that I'm in America driving a Humvee into DC, passing all of the monuments and people. We were driving on the streets, following one of the guys that knew the area. People and kids in the street were looking at us with their mouths agape, like, are we going to war? I'll never forget the looks on their faces, an older gentleman, many kids.

"It still amazes me that we were able to set up 5,000 troops that had all been flown in in just a few days. We go to wars and it takes longer to set up. And at the end of the day, no matter how exhausted I was, I still had to decompress every time, to just unwind from the crazy of it all before I went to sleep, or I would end up having the worst nightmares."

—*Mike, Airfield Management*

Friday, June 5, 2020

Escort Service

No Curfew

June 5, 2020 CMR 09-20 Secret Service Statement on Closures around the White House Washington D.C. –
The U.S. Secret Service, in coordination with the U.S. Park Police, is announcing the closure of the areas in and around the White House complex. These closures are in an effort to maintain the necessary security measures surrounding the White House complex, while also allowing for peaceful demonstration. Security fencing has been erected and the areas are clearly marked. The areas, including the entire Ellipse and its side panels, roadways and sidewalks, E Street and its sidewalks between 15th and 17ths Streets, First Division Monument and State Place, Sherman Park and Hamilton Place, Pennsylvania Avenue between 15th and 17th streets, and all of Lafayette Park, will remain closed until June 10.[1]

In these past few days, I ended up learning a lot about Snapchat. I had the app on my phone as I told my oldest teenager that she could only get an account if I got one and was her 'friend'. You know, in an attempt to monitor her social media and be a responsible mom, but really, I had no idea what I was doing, nor did I really have a desire to learn Snapchat

---

1 https://www.secretservice.gov/data/press/releases/2020/20-JUN/Secret-Service-Did-Not-Use-Tear-Gas-in-Lafayette-Park.pdf

operations. When I got my account, I figured out that Snapchat filters made me look 20 years younger and I could also take pictures of myself with cat ears and whiskers. As I never really had the need for pictures of myself with cat ears, I did not use Snapchat all that often. Or ever. Although the fountain-of-youth filters were lovely, I did not really see the point of looking amazing on social media, all to be a scary surprise in person.

After I saw the crowds and parades of people filming us as they passed, I started asking friends, and my daughter more about it, as I was pretty sure my face was published all over the Internet. I could not count how many times I had a phone camera shoved in my face as people marched by. In the end, I learned a pretty cool feature on it that I did end up using quite frequently. The same technology that was probably broadcasting my face was my direct link to viewing what was going on in the city. They were real-time videos that anyone could post at any location and it was then tied to a map. Where there were multiple posts, a 'hotspot' would develop. I could click on these and see the direct video from that location. At night on post, we would pull up various hot spots around the city to check in on how our friends might be doing and to see if we may be pulled into another location for backup. In the mornings before I went into work, I would look to see what was going on in the city before I even got there.

In the late morning of Friday, June 5, I pulled up Snapchat, tapped on Lafayette Square, and watched videos of people in the streets graffitiing one of the main roads with big yellow rollers. They were doing an incredible job and I was impressed at how professional looking it was. Then as the videos played out, I read gigantic letters that were painted from one side of the avenue to the other and down the street block, "BLACK LIVES MATTER."

Oh man, that is going to be a lot of work to clean all of that up.

And then I read the news stories. Mayor Bowser had granted permission for this to be painted, along with changing the name down a two-block section of 16th Street to 'Black Lives Matter Plaza.' It seemed to happen overnight, without warning. *Can they do that? Can someone just change a street name because they decided to?* Apparently, the Mayor can.

* * *

I watched more videos from the day before, before the storms had come in. The crowds did not appear physically violent, but they were screaming at the various federal forces that lined the park. There was a young man on a pedestal of some sort leading the chants. "They won't put down their guns and fight with us!"

"Yeah!" The crowds responded back, some raising their fists in the air.

"And if you're not fighting with us, you're fighting against us!" The anger that infused the city was unnerving.

I thought about the man that was leading the chants as I drove into the city that morning. I was trying to understand how they viewed us as the military. Some said that we were a good, unarmed barrier between the citizens and the police. Others associated us with the police. Others straight out asked us why we were there. It was a perfect question. And sometimes I would forget what the answer was.

I was stopped at a stoplight, lost in thought when a thin man, probably in his fifties, came up to my truck. I jumped, as he caught me off guard and was right next to my window, standing in the street. His clothes were torn and shabby. His eyes did not look right. They were glazed and he did not make eye contact. It was as if he was looking through me. I could not tell if it was because of him having a hard time looking into my window reflection or not. He swayed as he stood there. I instinctively pressed my door lock button to reassure myself that it was locked. I turned and looked forward at the light, praying for it to turn green. There were two cars in front of me and a line of cars behind me. I could not run the light, even if I had wanted to. I was trapped. The man was saying something, but I could not make out what it was. I just kept looking forward now, hoping he would move on and go away. I prayed he was just asking for money and not full of anger. In the past, I had given a few dollars, if I had it on me, to beggars at intersections, but this time it was different. I was in my uniform, so I did not want to take the chance, unsure of his intentions.

Then suddenly, he raised his fist and knocked on my window and waved his other arm in the air. I could see there was not a gun in them, but I was only slightly reassured by that observation. He was still right next to my face with only a piece of glass dividing us. He was still speaking loudly and not making sense. I felt my stomach drop. *Please turn green, please turn green,* I whispered with urgency at the light. Finally, after what felt like the longest moment of my life, the light switched to green, and I sped away as fast as I could without ramming the car in front of me.

I was shaking as I navigated down the rest of the streets to the Armory. *Why am I so scared of someone asking for money in broad daylight?*

What is happening to my head? I had never felt like this. But for the first time, I felt so hated—and in my own country. I thought about what the Commander had told us about the girl that was assaulted the day before. I was beginning to realize that I just did not know what people were capable of anymore. And I hated the feeling.

* * *

When I arrived at the Armory, we were gathered into our smaller groups for an informal briefing. Ripper came and spoke to our group. "Hey guys. We know that you've been working really hard the past week, and now that we have more troops here, we really want to give you a break."

Oh hell yeah. He was sending us home. We would finally get a day off.

"So, we're taking you off of the streets so you can rest here at the Armory and be on some lighter duty. We're going to divide you up to take different jobs like dispatch, escort service, meal duty..."

He continued on, but I stopped listening. *Meal duty? What the hell? We were just going to sit here and pass out meals for 12 hours? If I wasn't going to be manning the streets, I certainly didn't want to just sit around here.*

"Sir," I caught Ripper after he was done making his announcements. "We're fine to go back out there. We've been trained now and know what to expect. These new guys don't know, and we are going to have to waste so much time getting up to speed."

"Sorry, Sergeant. These are the orders and they come down from a much higher force than me."

I sighed. "Okay, Sir." I was getting so frustrated. I had been trying to be positive, trying to get through this week and now the shit sandwiches were just disguised in a different flavor. We were told to just sit tight in the bleachers until directed otherwise. And as much as I tried to be patient, my emotions continued to boil up. I had a family at home that I had not seen. I was missing important things at my 'day' job that I would eventually be so far behind in. If there were jobs for me to do here, fine, let me do them. If not, please do not make me sit around and wait in the bleachers for 12-hour shifts.

* * *

After excessively asking my Chalk leader to switch me from Dispatch to Escort Service so I could at least *leave* the Armory walls during the 12 hours that I was going to be stuck there, I realized that I was probably becoming quite a pain in the ass to them. I knew I needed to just suck it up and do whatever they told me to do, like a good airman would. But I think I was just bad at being in the military sometimes. After being out for fifteen years, and then only a 'weekend warrior' for the past three years, I think I got used to the comfortable life (i.e. normal life) and forgot how painful being on military orders actually could be. As the saying goes, I really just needed to "embrace the suck."

I was on Escort Duty with about fifteen other squadron members—which meant I would be escorting the bus drivers, who were contracted civilians, in and out of checkpoints and on and off the bases. They needed someone with a military ID to ride along and do this as they delivered troops to and from the Armory. This created much sit-around-and-wait time for us. If only I had brought in my computer, I could have at least gotten some work done. All I had at this time was my phone to scroll away on. I despised wasted time, so I think that is why I had such a difficult time with this.

"Attention in the Armory," the loudspeaker rang out over the arena. "South Carolina National Guard's Senator, Lindsey Graham, has decided to visit us today to check on his troops. Mr. Graham has also bought pizza for everyone here, so please help yourself at the far tables."

When the crowd cleared, I could see what must have been twenty tables lined up with Domino's Pizza stacked five or so high along the entire length. *Did this man seriously buy us all pizza?* I knew they had estimates of 3,000 or so Guardsmen there at this point. Granted, they were not all on duty or here at this time, but still, I was so impressed. Clearly this was a man that really was looking out for his troops. Well, all of the troops, really. It was nice to see a politician walk the walk. Senator Lindsey Graham did this. And no, I was not paid by his party to say this. But I think it is important for people to know these types of stories, too. Much like the goodie bags of happiness, these are the gestures that fuel the military's spirits. They do not go unnoticed.

I also had to laugh at the thought of the poor souls at Domino's that had taken, cooked, and delivered that incredible order.

* * *

We also had a few other visitors by this time. Barbara M. Barrett, the Secretary of the Air Force, stopped in to give a quick thank you speech to those of us that were in the stands at that time.

Then the Chief Master Sergeant of the Air Force, Kaleth Wright, stopped in to check on the troops as well. He also gave a thank you speech and then sat down with some of the guardsmen in metal chairs out on the floor where they had been sitting. He encouraged and welcomed anyone that wanted to just come up and join in the conversation.

"Have you met him before?" my MSgt asked me as we sat in the stands.

"No, I haven't."

"You should go up and talk to him. He really likes to meet everyone. He's a really nice guy. I've met him a few times."

My very good friend Roland, who was in the Baltimore Air Guards, had also met him and said the same thing about him. However, I suddenly had this vision of me going down to talk to him and tripping down the bleachers and landing at his feet. It just seemed like something I would do. "Ummm..." I hesitated.

"Just go do it," MSgt told me.

"Oh fine," I said. She was so bossy sometimes. But I began moseying down towards him, trying to look casual, as if I just accidentally would bump into him or something.

By this time, he was standing in a group of about four senior enlisted guards, listening to one of the ladies speak about her concerns in the squadron with her career field. I casually joined their circle, trying to make it look like I belonged there, nodding at the appropriate times when everyone else did, or when she made eye contact with me. Honestly, I only heard about half of everything she was saying. I was more interested in the Chief and how he seemed to be listening to her intently.

Finally, when she was satisfied with their conversation and appeared done, he looked over at me and asked, "How are you doing with everything going on?"

"Oh, I'm doing fine, Chief. It's been quite an experience, that's for sure."

"I'll bet it has been. But you're doing okay?"

You know how everybody asks you, 'how are you doing' all day long and it is more of a greeting than a question. The standard expected response is 'fine,' or 'good.' But usually people do not really expect more than that. It took me a second to realize that he actually was asking. He seemed to genuinely care about how everyone was doing.

"Yes, it has been challenging, I suppose, in some ways. I personally just don't think I was really prepared mentally for what we were going to experience. But I'm not sure that anyone knew exactly what to expect. This has just been so different from anything we've ever done."

"I think you hit the nail on the head there, Sergeant—Py-rah, is it?"

"Yes, Sir, that's right." I answered him, throwing in a 'Sir,' because in the Air Force they call their enlisted 'Sir' or 'Ma'am.' That is a big no-no in the Navy; you only call the officers Sir, or you will be met with a "Don't call me sir, I work for a living." It was one of the weirdest differences to get used to. But the Air Force called anyone "sir" or "ma'am" out of respect. I could get on board with that. Eventually.

"Well, you let me know if there is anything I can do to help you, Sergeant. Just send me an email."

"Okay, well thank you so much."

"My pleasure. Take care now." He pointed his elbow at me, to which I respectfully touched my elbow to his. It was this strange new alien-like greeting that the year 2020 invented in an attempt to stop the pandemic spread of germs through the standard 'shaking hands' greeting.

We took a quick picture together, which was a really cool keepsake. I later sat down and composed a long letter of all of my complaints about the Air Force and sent it off to him. Just kidding, that would have been ridiculous. A little lesson to anyone new to the military: do not send the email that they tell you to go ahead and send to them. Lies. All lies. And I was not about to fall for his little trick.

* * *

"Does anyone here have good handwriting?" One of the Intel Master Sergeants called from the bottom of the bleachers. I looked up, scanning to see if anyone had raised his or her hand.

"Anyone?" he asked again.

It just so happened that I did possess the secret power of amazing handwriting, but I was never sure whom I could trust with that kind of information. Since Intel guy had been standing there for a while with no takers, I felt like I should at least offer to help him out. Slowly I raised my hand about halfway up and kind of shrugged. He pointed at me and I replied, "It's pretty good." Which was a lie because my penmanship was amazing, even though I could not direct traffic. See—know what you are good at. (Also, I'm good at eating tacos and drinking wine, just in case we were keeping track.)

"Alright great," he replied with enthusiasm. "We need some help up in the office, and hey, at least you'll get to sit in some air-conditioning." It was like he knew the way to my heart.

Seduced by the words *air-conditioning*, I loaded up with all of my gear and wearing my bulletproof vest, I followed him, excited to at last be busy. We went up the stairs and through the doors where the sweet, cool air of delightfulness smacked me in the face.

When we arrived in the office, it was full of many high-ranking officers colliding in an attempt to organize the chaos of this unprecedented situation. The Chief Master Sergeant of our squadron turned towards us as we walked into the office. "Handwriting?" she asked.

"Yes, this is Staff Sergeant Pyrah who can help us with the documents." Intel informed her.

"Oh, yeah, we don't need her anymore. We figured it out." And as quickly as I came, I was dismissed.

Well, I guess that burned another fifteen minutes or so of the day. Intel looked at me, "I'm so sorry to drag you all the way up here." His face told me that he was genuinely sorry.

"Oh, it's totally fine. It's not like I was doing anything. I welcome the change of scenery," I reassured him.

"Well here, at least come and sit in my office and cool down for a while." I went into his office and I spent the next fifteen minutes or so catching up and letting him know what was going on out on the streets, while sipping cocktails and feeling the cool air pleasantly bring my core temperature to a

Friday, June 5, 2020

satisfying degree. And when I say cocktails, I really mean a bottle of Deer Park water, but it tasted just as refreshing in that moment. At least that is what I convinced myself.

After we had wrapped up our visit, I had nothing left to do but return to the bleachers. I had to admit, I was a changed woman after seeing how the well-offs were residing. And I could not help but be a little reluctant to take the steps back down the stairs to hang out with the rest of the peasants. As I was about to exit the cool air, the doors opened abruptly and none other than my favorite gray-haired general came walking in with his entourage.

"Sir, how are you?" I was so excited to see him again.

"Hi! Doing well. How are you doing?" the General asked me.

"I'm doing well… as well," I replied, cringing, as I heard what I had just said. I was not quite sure if he remembered me from the other night. I am sure he had seen thousands of troops since then.

"No more bricks in the street, I assume?"

"Oh, you remembered! No, Sir. No more bricks."

"Of course, I remember. You're the writer, correct," he asked me.

"Well yes, I am."

"I can't wait to hear how this story turns out." He turned towards the door. "Well, you have a good day now." And with that, he and his group quickly retreated down the hall.

"Thank you, Sir. You have a great day, as well," I replied, raising my voice as he walked away so he could hear.

* * *

I went back down and found my place on the bleachers next to my new friends in the squadron: the weather guy and the SERE (Survival, Evasion, Resistance, and Escape) School instructor that I spent much time during our day training in Ellipse Park exchanging SERE School stories. I had previously gone through SERE school during my time in the Navy, so it was fun to compare crazy stories. This was indeed a highlight of the week. I was able to meet and work with many new faces that under our normal squadron life of drilling once a month, and working mostly in our own respective shops, I would not have had the opportunity to get to know. We spent much of the afternoon waiting for an assignment and making up stories with our play on words now that we were running an "Escort Service." I was given the callsign 'Butch 001' and the other two were 'Big Daddy' and 'Little Bro'—which still just seems weird, so I'm going to call him 'Weather' for the sake of this book. Because Big Daddy is not weird at all.

"Now I expect you to respond over the radios, 'Roger Big Daddy,' anytime you confirm coordination with me. Got it?"

"Got it," I answered him, trying not to laugh. I was beginning to see how great the afternoon was not having someone yell derogatory words at me, and out of the middle of political battles. Maybe I needed this mental break from the streets—a change of scenery, if you will—more than I realized.

* * *

"Alright, I need three escorts," one of the dispatchers announced to us waiting in the bleachers.

The three of us could not have stood up faster if the seat had been on fire. "We'll do it! Whatever it is!" We just wanted to get out of there more than anything.

We were instructed to take three buses and get them over to Joint Base Andrews, as there was a C-130 landing soon with a plane full of more Army National Guardsmen, this time from Indiana. There were still guardsmen coming in. I could not believe how many people were being called up, especially as things seemed like they were beginning to wind down.

We arrived at the Andrews Base theater sometime after 2200 (10pm) and waited. And then waited some more. The bus drivers and escorts congregated in front of the theater, under the overhang. It was a quiet night, still hot and humid, but since we did not have to carry any gear or walk far, it was quite manageable and not too uncomfortable. Somehow just being in our long-sleeve uniforms and combat boots on an 85-degree night did not seem bad at all anymore. It was all about perspective.

At last the aircraft landed and the troops were delivered to our bus's location on base. Big Daddy's and Weather's bus had already filled up and they took off for the Armory. Soon the rest of the troops began to climb in mine. They were beat. I could see it in their faces. Apparently, they had been up for 24 hours now, waiting on all of their travel and activation processes and paperwork to fall into place. They had been activated, then told to stand down and go home, and then reactivated. It had been a mess.

My bus was loaded about halfway, and I walked out to the dispatchers who were directing and coordinating the buses from the theater back to the Armory. "I'm about half full, is there anyone else on the C-130 that was coming back to the Armory with us?"

"No, this is everyone from this plane. However, we have another aircraft headed this way with more troops, so maybe stick around," the dispatcher informed me.

"What is their ETA (estimated time of arrival)," I asked him.

"Not sure. We're just told they are coming."

"So… they haven't landed yet and you have no idea what time they are landing?"

"Yes ma'am," he replied to me.

I was suddenly having flashbacks to two nights ago, waiting on the streets into the wee hours of the morning for the bus to come and pick us up. "Oh, no. That's not going to work. I will send another bus out here, but these guys have been up for 24 hours and I am not going to make them wait on a plane that you don't even know when it is scheduled to land."

"Um, okay, just send us another bus then," he replied to me slowly. I knew I was projecting my anger of the chaotic week on this poor guy, but I was just so fed up with junior airmen and soldiers being jacked around unnecessarily. There were so many moving parts to this "spring to action" activation of 5,000 troops in three days. Wars take longer to set up. And the disarray of the situation was happening from the very top, as Army and Air Force leadership from all over the country were caught in an extraordinary drill, working endlessly to create some sense of order.

* * *

"I equated a good portion of the domestic operations to the buildup of a deployment. From the start, there was an anticipation being built up that we were going to a hostile environment. The call was quick, a recall of less than 24 hours to report to the armory. Everything bled together, from issuing of equipment, to the so called "riot control" training that we received. A disorganized chaos of personnel having no idea where to go, who their tactical level leadership was, and even who was going to feed them. So many questions were being thrown around, and no answers were being given.

"Typical days started around 1400 (2pm) and ended around 0400 (4am) the following morning. Lunches were served out of a plastic box and felt like leftover Subway. A large percentage of the day was spent waiting for something

bad to happen so that the quick reaction force would be forward staged. When the team did get sent downtown, it consisted of sitting around during daylight hours and prepping for after curfew hours. The duties seemed to be picking our noses at memorials or guessing which federal agency was better equipped.

"The waiting was the worst part. Sometimes the waiting resorted to complaining fests. Other times it led to wild conspiracy theories, wrestling around. You learn so much about the people you are serving beside just because there is nothing but time. Even while waiting in the rear behind the main protest line, we would talk over the radio about when we spotted beautiful women or laughing at some of the signs that were made. At one point in time some told us to "Fuck our heart." I still am wondering the context to that statement. In no way am I attempting to belittle the movement, or make light of what is happening in society.

"Sometimes when you go through stressful situations, you have to find laughter in all of the chaos. One of the last nights during a thunderstorm, a whole group of people were huddled by the National Christmas Tree and surrounding buildings. I looked down and thought I saw marbles sitting next to a curb. I knelt down to pick it up and it felt soft. I had no idea what it was, and my curiosity got the best of me. I put my tongue to the "marble," and my sinuses immediately went crazy. I just tasted a pepper ball that had come out of a hopper from a park service paintball gun. Everyone got a good laugh out of it, and for a split second, everyone stopped thinking about how miserable and wet we were. When the team did get sent downtown, it consisted of sitting around during daylight hours and prepping for after curfew hours.

"The typical team was small, 15-20 people, consisting of personnel that were never trained—or ever should be—for such a situation. The bigger implication was whether people wanted to be serving their country in this capacity. Some of my Airmen even questioned their moral compass when they saw their friends and high school classmates on the other side of protests. The camaraderie that was built between service members, however, was somewhat reminiscent of a deployment. Some of the things that we bonded over, similar to that of a deployment, were things that only people that serve together share. Whether it was laughing at protesters throwing bottles filled with piss and hitting trees above their head. Or a full gallon jug of milk crashing into a riot shield. Only to be laughed at by someone making the joke, "as long as it isn't that hippy skim milk shit."

"While actively engaging with protesters, there were a multitude of emotions circulating in the atmosphere. The entire situation was dependent on where

Friday, June 5, 2020

you were working and who you were working with. Lafayette Square seemed to be the epicenter of protesters and all walks of federal agencies. On any given night, they would position members of the National Guard between police and federal agents between the protesters. The situation would be calm until one side agitated another. A barrage of frozen water bottles, bricks, or even old CS canisters (tear gas) would incite the authorities to "move the line forward." This would further the escalation between authorities and protesters, which made me personally feel like the mediator in a bitter divorce. Most protesters had no ill intention towards members of the National Guard—to the point that protesters would explicitly tell us that they knew we were only doing our job.

"The entire series of events are unable to be captured in any form of media. During any given night, media outlets would show up for an hour or two. They would never capture the waves of nonviolent to violent protesters. Some of the things that were said, the actions that both sides took, can never be justified. It is hard to take a position on either side, being a military member while also being a supporter of civil rights. Some people will say that taking no action is being complicit. I would say that people need to stop standing behind the voice of corporations or a movement and start being human beings."

— Alek/Weather

Saturday, June 6, 2020

Weekend Vibes

"Mold skin?"

"No, *mole* skin," MSgt said as she emphasized the letters.

"Um, that sounds terrible and I don't think I'm interested. At all."

"It's not *really* mole skin," she laughed. "That's just what it's called. And it can really help with your blister."

"That just sounds so disgusting." I turned around. "But I will try it. Anything." I limped over to the table where the Army nurse was set up.

I stepped up to her table. "So, ah, I heard you have mole skin." It was the weirdest thing I think I have ever said to anyone before.

"Oh yes, we do. Do you have a blister?" The nurse asked me, with a concerned look on her face.

"I do. I heard that could help?" I still was not convinced.

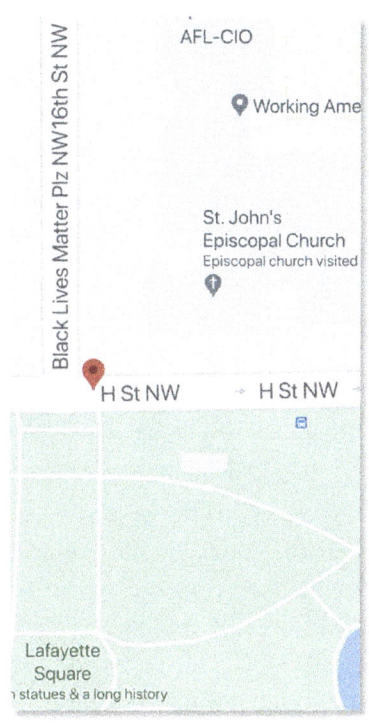

"Oh, this will really help. Where's your blister? Here, why don't you take off your boot."

I did as she instructed to reveal the demon parasite that was chewing through my bare skin. The back of my ankle was red, angry, and bubbled up to an unusual height.

"Well that is a good one. Here, we'll fix you right up." And with that she went to town on the mole skin, which really did not look like mole skin. It was sticky on one side and soft and almost furry like on the other. Okay, yes, I realize that sounds a lot like what mole skin would look like. After her delicate patchwork, I was finally able to slip my boot on and no longer walk like I had a peg leg (to avoid bending at the ankle).

As soon as I got my boot back on, I heard the dispatcher yell my name.

"Sergeant Pyrah, I need you to go into Lafayette Park and pick up the troops stationed along the streets. Now."

"Yes, Sir," I'm on it.

It was Saturday. I was doing another day of escorting. Today was also supposed to be the day of the Million-Man March: an attempt to recreate Dr. Martin Luther King Jr.'s Million-Man March. We were projected to have a million people marching throughout the city. As a result of this, we were extra staffed up. The Armory was bustling with Guardsmen, some getting their riot gear prepped. Some were second string, prepping if backup was needed. And some were just there like me, eating free Girl Scout cookies until called upon to make a bus run.

As I climbed into the bus, I was greeted by the sweetest little lady, probably in her mid-sixties. She had wavy black hair and was so petite that her feet barely reached the gas pedal. She reminded me of my mother, as she had a very nurturing way about her. I sat in the front seat of my air-conditioned bus, chatting away with the bus driver, Doris, as we made our way into the city to pick up some troops that were nearing the end of their shift.

I was so relieved that she seemed to really know her way around the city, because I had no idea where we were going. As we came into the checkpoints on the streets where the police barricades were blocking off the usual traffic, she would turn on the inside bus light so the policemen could see me in uniform, holding my ID and often just waved us through. However, I could not just stand there holding my badge like a disgruntled drill sergeant. I flashed it at the cops with flare, as if my military ID was some type of VIP pass to get me into the secret society of the streets.

Saturday, June 6, 2020

But as we got closer to the Park, the crowds began to grow. They were not just limited to the sidewalk; they were also filtered into the streets.

While we drove, I was just taking it all in. I could not believe how many people were out. Now granted, it was not a million. However, it had to be a couple of thousand. But Doris seemed to grow more nervous. Her eyes were glued to the people in front of us and to the sides. She became extra vigilant. She insisted that we proceed much more cautiously. "You just sit down. I'm going to keep the lights off, so they don't see you."

She was actually shutting the bus lights off, so they did not see me standing up inside the bus in my military uniform. It was a very peculiar feeling. I realized that she felt like she had to protect me. And it was not until then that it started to make me nervous—*am I missing something? Should I be worried too?*

I shook my head. "Oh, it's okay ma'am. This is what I'm here to do." But she continued to insist that we keep the lights off as we approached. As the bus crept along towards the White House, the crowds also grew louder as they chanted. And then, as it unveiled itself before us, both of our jaws dropped.

"Does that really say..." her voice trailed off.

"Yep. It sure does," I answered her. There, in gigantic, bright orange letters written across the street were the words, "DEFUND THE POLICE." A chill ran up my spine. *Defund the police? And then what? Did they really want a city without policing? How would that help anything, especially the violence?* To me, the thought was terrifying. I could not understand what was happening to our country.

"I need you to sit down, Sweetie. Sit down right now so they won't see you," Doris spoke with authoritative urgency. The empty bus, empty except for the two of us, was now completely surrounded by the crowd. She was still able to drive, but at a snail's pace as we made our way down the street. And in this crazy twist of events, this little old African American woman was trying to protect *me*. Me—an Airman in the National Guard. And I was either too naïve to be scared or unwilling to allow myself to be scared, as I was supposed to be looking out for *her*. It was the strangest of times in so many ways.

Eventually we navigated through the crowd and down the street where we picked up several of our troops and brought them back to the Armory.

On my drive home from the Armory that night, I kept thinking about

Doris. I wondered if I would ever see her again. It is amazing the people that come into our lives for sometimes just a moment under the strangest circumstances, but the impact they leave behind will always remain.

* * *

"I received a call from my shop superintendent on Friday afternoon, letting me know that I had two hours to make it to the base with a 72-hour bag. Details of the mission would not be provided until I arrived due to the nature of the unique mission. I had just sat down to eat my dinner around 8 pm when I had to stop everything to get ready and head down to the base.

"Upon arrival at the base, everything looked surreal to me. I saw people being loaded into buses, fully geared up with riot gear, battle helmets and vests. Soon after I saw that I received my briefing. I then understood what the mission was about. During my briefing, the protest was projected on a large screen for us to see what was happening downtown. It looked pretty bad and it also mentally prepared us for the worst.

"On my first day, I was placed with a group of 54 airmen to work closely with the U.S. Marshals to protect the Washington monument. I guarded the Washington Monument for 18 hours, all while looking at the sky and amazed at all the helicopters flying and listening to the protesters going by. On days two and three I was placed with Guard a block with two other airmen in front of the World Bank. The blocks surrounding the World Bank were routes where the protesters would march by, since the White House is near. The other days our mission would change from day to day. We guarded the White House, patrolled the streets, and practiced with the RIOT Control team. As the days went by, I realized and I noticed that the situation in downtown D.C. was not as bad as the media presented it. In fact, the overwhelming military presence upset the protesters, which led to some of them insulting us and throwing bottles at us. Often, I would be asked why I was out there, who sent me, why were we doing this? The only answer I could give was that, "I swore an oath to serve and protect" and that we were there to do a mission and protect everyone.

"It was upsetting to me to realize that the protesters thought we had weapons, they felt threatened by the military. Also, the media mentioned that the military had weapons, and so did the President. Police and those who are required to carry as part of their job had them. We did not have any at all, and I was glad because the military's mission is a lot different from law enforcement.

Saturday, June 6, 2020

Overall, it was two weeks of intense physical and mental exhaustion. I was insulted, we had items thrown at us, and stood for hours in 90-degree heat. It was not fun, but we had a job to do. What I saw downtown was not what the media projected. There were lots of protesters, mostly peaceful, some upset because of the overwhelming amount of law enforcement, in addition to the military presence. Overall, in the end it was a historic experience, which I hope does not repeat."

— *Louis, Aviation Flight Equipment*

Tuesday, June 9, 2020

L Street & 11th

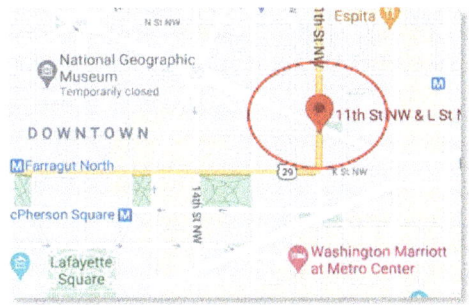

"The mission has changed. Meet at JBAB 1400," was the message that we woke up to this morning.

We all met at a designated building at Joint Base Anacostia-Bolling in Southeast Washington, D.C. Our assignment today was to work directly with the Army and drive tactical vehicles in different checkpoints across the city. As the rioting had really begun to calm, the other Guardsmen that had been brought in from across the country were slowly being sent home. D.C. was once again being guarded by just the local National Guards—Army and Air Guards of D.C. Since our numbers were decreasing, there were also not as many posted to a checkpoint. We now only had two or three military, and two or three federal agency members. We were not directing traffic anymore or blocking the streets, but instead we were a presence, instructed to report any encounters of tension.

After we were briefed of the day's mission, they sent our Chalk to the backside of the building where all of the Army's tactical vehicles were lined up. Before they could assign us where to go, I scurried past the line of Humvees over to the most impressive looking truck I could find. It was another LMTV. There were only two of them in the long line of the armored Humvees. The LMTV sat high, allowing its riders to sit high, with great visibility all around.

"Hey!" I called out to the driver. He was a short soldier with the most striking hazel colored eyes against his brown skin. Campbell was Jamaican and could not have been more than nineteen years old. "Do you care if I ride with you?"

"Fine with me," he called out over the idling engine noises. "You'll love this one. I picked it because I knew it had air-conditioning." I knew at that moment that Campbell was my spirit animal.

He gave me a thumbs-up and quick tutorial of how to climb up the steps into the massive vehicle, and how to work the door. Although it sounds like something that would be very elementary, the door handle had a secret safety lock from the inside to keep dangerous people out during wartime, and to publicly erode my self-esteem during domestic operations. Apparently, it was possible to open it from the inside too, but only if one is a professional wrestler. I clearly was not, and after attempting and failing several times to open it on my own, I made a mental note to myself that a few extra bicep curls a day would not do me any harm so I did not end up having to ask for help to get out of an LMTV again. You know, for the next time I found myself in one.

On the inside, there were two seats, for a driver and passenger, and then a tiny little shelf-like seat made for a leprechaun or a man with legs that were only two feet long. The shelf seat was located back and center where, due to being the shortest (although my legs were at least a good three feet long), I was the lucky winner that got to sit with my feet sticking straight out towards the windshield. But it was all good in the hood, because I had air-conditioning and a great view!

This LMTV was painted a fashionable desert-sand color (vs. the green one we had the previous week) and the back end was like a covered wagon. When I later climbed into the back, I imagined this is what they must have felt like on the Oregon Trail. And then I went out and shot a buffalo, but only could drag back 99 pounds of meat, because taking two trips was not a possible solution.

As we rode down the streets of DC from JBAB to our post in our convoy of armored vehicles, we were either met with an excited wave or the angry middle finger. Nearly every other person had their cellphone in the air, recording us, no doubt with all kinds of subtitles being sent all over Snapchat. I tried to smile and look friendly, even flashing the peace sign at the cameras in Forrest Gump fashion, hoping that it would deter any negative messages that were being sent out across the world about the National Guard. Still, it was challenging to ignore the random shouting of "get out of my city," despite the fact that we were the D.C. Capital Guardians. This *was* our city, and we were only there under orders to protect it. Many of the personnel in our unit lived directly in the city. They were the ones that did not know how to hunt buffalo and left that to us country folks.

As soon as we parked the massive truck on the street edge, a new set of DEA guys gave us the SITREP (Situational Report). Most importantly, we were briefed where the nearest and cleanest bathroom was. Today, it happened to be right on this corner, as we were next to an air-conditioned, clean federal building.

They were no longer closing these side streets, as the number of people protesting was becoming much more manageable. We were there in case things turned ugly. And for the most part, it was a pretty quiet day, which was very welcome. In fact, we were instructed that we no longer needed to wear our bulletproof vests and our helmets (though we kept them with us in the truck). Essentially, we were just normal, unarmed people, in uniform on the streets of D.C. in an armored vehicle convoy. So *mostly* normal.

After we parked, I had to get out and walk around a bit. Eventually, I ended up sitting in the covered, but open back end of the truck, as it was facing the street and was a great platform to just hang out on. Because it was probably about six feet off of the ground, it even felt a bit safer in there, as I sat by myself. The other two guys remained in the pleasant, air-conditioned truck, which would have been more comfortable, but I was getting bored sitting there. The back end was hot, but it had a great view of everything. And because there was not anything going on in that moment, I took advantage of the time and even dialed into a couple work meetings from my 'day job'.

I propped my phone up on speaker, and as I watched people pass by, I was also able to listen in on the latest update to the delivery schedule for training devices, just like I would do on any other Tuesday afternoon. I would begin to think everything was back to normal, and then someone would drive by and yell "Fuck the Police! Fuck you all!" out of their car window, like any good citizen, and it would quickly snap me back to reality as I scrambled to make sure my phone was on mute. Tensions were still present.

* * *

During one of the quieter moments, I started to chat with Campbell. He told me that he had been sent in on that first Friday, May 29th when things started to get crazy. They were part of the group that was voluntarily activated (meaning he volunteered) to aid in Domestic Operations with the COVID-19 Operations. He said they had finished up duty Thursday night and were called into base early Friday morning, as they were already on orders and it would be easy to send them in as things were beginning to heat up.

"Friday was bad. Somewhere there was a pallet of bricks" [the jury was still out whether it was brought in specifically for destruction, as rumor had it, or it was left on a job site]. "We were all lined up in front of the White House, as the Secret Service and Police needed backup to form a solid barrier.

"They just kept throwing bricks at us. Most of them bounced off of our shields, but some made it over. One of our guys got hit in the head and last I heard had a severe concussion. Another guy's leg got hit and was all messed up.

"Someone threw his skateboard at us and then some dude actually threw his bicycle at us," he said to me, laughing. "What kind of an idiot throws his own bicycle?"

I was dumbfounded listening to his stories. I kept asking, "are you serious?" He just nodded. "Were you scared?"

"Nah. I didn't care. But there was this short girl next to me, she couldn't have been more than 4-½ feet tall. She was just quiet and shaking the whole time. I kept asking her if she was okay, and she just nodded, her eyes never leaving the crowd in front of her, and just shaking.

"But you know that church, that one that they set on fire?"

I nodded, hanging on his every word.

"Earlier in the night, there was a pastor from there that came out. She went down the entire line and stood in front of each one of us and said an individual prayer for everyone. I'm not super-religious or anything, but I think her prayers may have kept us alive that night," he shrugged as if it was just any other day at work.

* * *

As the afternoon sun began to set, a group of four teenage boys walked across the street in front of me. They were dressed in baggy pants that barely clung to their body, leaving their underwear hanging out of the top. I wanted to go and pull up their pants and hand them a belt, like I would with my own son if he came out of his room dressed like that.

The first one came up to the back of the truck where I was sitting alone, the bed of the truck coming up to his neck. "Yo, you got any water?"

There was a pack of water in there. I'm not sure how long it had been there. Maybe a day or so.

"Umm, I do... but it's warm."

He shrugged and stuck out his hand.

Seriously? I would not let my teenager speak to me in that manner. If I was going to give this boy any water, he was going to ask me respectfully.

"Okay... what do you say?"

He narrowed his eyes, no doubt thinking, did she really just ask me that? For a second there I was not sure if he was going to hit me or laugh at me. But I just looked at him with my most stern mom look. I was not budging. I couldn't.

"Ummm, please?" He slowly formed the word, as if it was almost painful.

"Sure," I replied with enthusiasm and handed him the water. And they subsequently went down the line, each asking with a please and ending with a 'thank you.'

"You're very welcome."

I am sure they were annoyed, but as a mom, it felt like one of those small victories. And I could only hope that if my son were to ask another mom for something that she would ensure that he spoke properly and used his best manners. Because that is what moms are supposed to do for each other.

* * *

There were not many protesters going by at this location. Every now and then we would see a small group with a sign or two, as if they had just come from a protest somewhere else.

None struck me more entertaining that day than a group of teenage girls that came by giggling, as if they were on their way to a slumber party. In their hands, they carried various signs. The blonde girl in a pink miniskirt had a sign with so much writing on it, I had to ask her to stop so I could read it all.

"Wait, what does your sign say," I asked, intrigued, as I had not seen one that was so full of print. She faced it towards me, and I read the words out loud, "DONALD TRUMP YOU'RE A SLUMLORD BULLY WITH A BANKRUPT SOUL. YEAH YOU'RE A BUM-FACED BOZO WITH TINY HANDS AND A CANCELED SHOW."

I realized my facial expression was saying judgmental things and quickly corrected it. "Ummm..." I really did not know what to say. *That was the sign she put together to take to a Black Lives Matter protest? That is the sign she wanted crowds, possibly even news reporters to look at and know that she was the one that took the time to stencil that in?*

"Um, yeah, it's literally a TikTok song. You're probably too old to know what that is." She flipped her hair over her shoulder as she said it and they all giggled again and kept making their way down the street. When they were out of sight, I turned around and shook my head. And while I did know what TikTok was, she was right about something. I was beginning to feel way too old for this.

* * *

After the giggly teeny-boppers left, we stood around on the shaded corner and visited more, talking about where everyone was from, their kids, the usual things. As we were talking, I watched this little old Indian woman walk gently down the sidewalk towards us. She wore a simple but colorful dress and even had a bhindi on her forehead. She stopped in front of me and I prepared myself, never quite knowing how people were going to react to us.

"I prayed when you go to wars, and now I pray for you here," she spoke and took my hand in hers. She smelled like flowers. During this unusual time, where the entire world is so conscientious as to not spread germs, she did not even hesitate to touch a stranger. And while we were all supposed to care so much about germs and not touch anyone, it felt so nice to feel the touch of kindness from a stranger, a touch that matched her kind words.

"Well thank you so much. We appreciate your prayers and kind words,"

Tuesday, June 9, 2020

I said back to her.

She looked me in the eyes, "I watched on the TV, and I prayed for your safety, because you keep us safe." She smiled and said, "bless you, bless you all." She turned away and continued to make her way down the street. And I smiled, my faith in humanity restored, and I was reminded why I was out here.

* * *

"While religiously watching the civil unrest in DC on the nightly news, I concluded it was only a matter of time before we were called to help Metro PD. In fact, my commander prepared me for this possibility via a message through a popular professional networking app. Every member of the unit is to be activated and report to the D.C. Armory, or as we call it, Joint Force Headquarters. The best line from that message was "stop drinking so you can support coherently." Our commanders knew COVID-19 restrictions forced all of us to telework, which, let's be honest, meant we were doing about two hours of real work per week and becoming borderline raging alcoholics. Luckily, when I received the message, I was sober. I grabbed my gear and raced to the armory, as did everyone else in my unit with a shared curiosity as to what we would be doing. Upon arrival, I was told to sit in some bleachers until everyone was there. The epitome of 'hurry up and wait.'

"I naturally chose to sit with one of my closest friends, the Staff Weather Officer. We immediately talked about plans of consulting leaders with our expertise, so the mission is successful. The both of us even approached the appointed Task Force Commander but were somewhat shot down. I wanted to develop a plan to facilitate personnel recovery in case our people got overrun with angry protesters, and my weather buddy wanted to provide daily technical weather information.

"After receiving about thirty minutes of riot control training, getting deputized and issued gear, I was unconfidently ready to protect federal monuments. The first night consisted of about 15 of us deployed to protect the World War II memorial. By the time my small team was bused there, you could easily see previous day protester desecration. Washed-off spray paint and vandalism had already occurred which made me think, "who the fuck would do this to the WWII memorial?" Regardless, we were there to protect and enforce the Mayor's declared 7 o'clock curfew, resulting in an uneventful night. I did receive a hit of reality while on this shift though. I could see a small pile of bricks stacked against a light post near the south entrance of the memorial. These were not the red bricks you are accustomed to; these were solid pressed grey concrete. After further inquiry with our Intelligence folks, I found out extreme agitators paid to have pallets of bricks forklifted to strategic locations to throw at us and the police. That is what sparked a clear distinction between peaceful protesters, who genuinely believe in change, and those who are funded and organized to incite violence.

"The second night, my team and I were redirected to Lafayette Square. This is where a large group of us spent the next 30 days or so, primarily to stop an occupation of The White House. You laugh, but yes, this was the goal of the protesters; peacefully occupy the White House long enough until the government caves to demands. Obviously, that did not happen. In fact, there was nothing peaceful about the attempt! All D.C. National Guard troops, federal police, and supporting police departments patriotically protecting Lafayette Square witnessed bricks, frozen water bottles, paint-filled balloons, and piss-filled bottles being tossed their way. Some of these professionals even got lasers pointed in their eyes and as you know, that can cause permanent eyesight damage.

"In one 'police line' I was in, an Air National Guard member received this laser treatment which triggered an epileptic seizure, something he had not experienced in years. Needless to say, the overall message of protesters was not clear. Even the rhetoric we heard was not clear. Occasionally we heard chants you undoubtedly heard on the news like 'I can't breathe!,' 'Hands up, don't shoot!,' and 'Black Lives Matter!' But once that got old, we were bombarded with screams about every national issue from abortion to how many genders there are. All of this sounds bad, and to any person not part of a profession of arms or law enforcement, it was. Again, I have nothing against anyone's peaceful right to protest, in fact, I encourage citizens to do so and I was proud to be there to protect that right. But violence seems to be more divisive.

"Something I never thought I would feel after this operation though... I secretly want to go back. I want to do it again. Not for reasons you may think though. Much like combat operations, Soldiers, Sailors, Airmen, and Marines who have deployed feel the same way. Experiencing chaos with those you would die for offers a beautiful glimmer of humanity that you will not see anywhere else, but you would do it again in a heartbeat."

— Miles, SERE Instructor (aka Big Daddy)

Wednesday, June 10, 2020

Day Off Down South

Wednesday, June 10th, was my first official day off in eleven days. I could not have been more excited to be out of uniform, and blister-causing boots, and just be a 'normal' person today. I attempted to sleep in, but 7AM was as far as that would go, as my body had become accustomed to function off little sleep at this point. My husband's alarm went off and I stayed in bed, chatting with him, as he got ready to go into work. We made a quick plan for dinner and he kissed me goodbye leaving me to the quiet house.

I continued to lie in bed for another hour, basking in the joy of not having to do anything today but spend time with my family and friends. Soon, my door gently opened, and my two daughters crept in the room and climbed in bed on either side of me. I listened to them catch me up on everything that they had been doing and all of the teenage drama that encircled their lives. It was wonderful to listen to their stories. It felt so good to hear of their summer plans, their worries of boys, and adventures with friends. It was a break from the hatred I saw in the streets. I was in my home, safe, and loved.

That afternoon, my friends took me out to an impromptu lunch to spend a few hours together. It was the first time I put 'real clothes' on to go outside in a week and a half and drove somewhere else other than a base. It was a wonderful feeling. We went to a restaurant on the water, as outdoor seating had just been reopened for local businesses for the first time since March, when everything was mandated to be shut down due to COVID-19.

We caught up on everything about work, and our families. As I listened, my thoughts kept drifting back to DC and what was going on up there. It was impossible not to bring up the stories in my conversations. It was what was weighing on me nonstop. Although my friends knew I did not want to think about it, they were also very curious about what was going on. As I retold some of the stories, it was challenging to describe them without conveying my own biases, but I knew the last thing I wanted to get into was anything political, so I did my best to just talk about how I felt. And I found that as much as I wanted to forget about the stories, I also needed to get them out of my head. It seemed that talking about it all was the best way to do that. I was safe down here on the water, in Southern Maryland. It did not look anything like the streets of DC, and that was very much welcomed.

When the waitress came back around, she asked if we wanted another drink.

"Oh, that would be lovely," I said with a smile, knowing I did not have anywhere I needed to be any time soon.

"No problem, Sweetie. But I do just want to let you all know, there is a protest scheduled to be here in a few hours, so you may want to be out of the area by then, or you could be stuck here for a while."

My heart dropped. "Are you serious?" I exclaimed without putting much thought behind my words. "A protest? Down here? Ugh. I can't escape this crap!"

I do not think the waitress really knew what to say, so she just ran off to get our drinks instead.

"Oh my god," I turned to my friends. "I have to get out of here before that happens. I just don't want to hear any more opinions being shoved down my throat today."

Suddenly my friend's twenty-something year-old daughter got up from the table, slammed her napkin down and walked away, refusing to make eye contact. It was apparent that something was wrong.

Oh crap. The entire time I was in the city, I had been keeping my opinions to myself and focusing on not reacting to the crowds. I was so tired of being 'on' all the time. I had let my guard down and spoke freely, knowing that my friends knew me enough to know where my heart is. They understood that I was tired and frustrated. But to some that may not know me as well, what I was projecting was my intolerance for the general ridiculousness that I had been putting up with on a daily basis. It was nice to vent but I needed to put my game face back on in preparation for tomorrow.

I had been insensitive, but I was unaware that I was offending someone. I think this is where both sides of opposition are failing each other. There is a growing reluctance to listen to others who may have a differing position. This was why we were in this place of 'us against them.' This was why our country has been so divided. This is what was perpetuating the hate and violence: misunderstanding and intolerance.

She was so young. The world had shaped her in a different way from her experiences, just as it had shaped me and continued to daily.

I believe someone pretty important once said, "we must learn to live together as brothers or perish together as fools." (Spoiler alert—it was MLK.)

Thursday, June 11, 2020

Back to the Streets

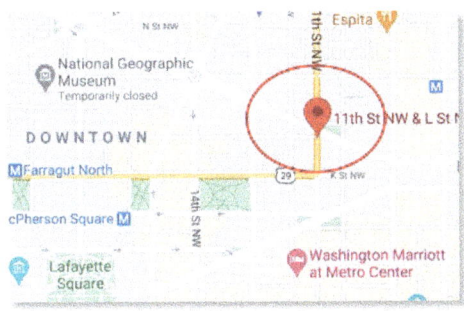

"Where are you?" My MSgt asked me over the phone in an urgent whisper.

"I'm about fifteen minutes out," I replied. "Traffic was stupid today for some reason."

"Well just hurry. Senior Master Sergeant is not happy."

"Okay, I'm sor-reee." I hung up and broke the speed limit some more to get to the base even faster. Every day it seemed we waited around for at least an hour before we briefed and proceeded with the plan of the day. The one day that I am running late, we were mustering right on time. I began chewing my nails and turned up my car radio to a ludicrous volume to try to block out my anxiousness.

When I arrived, I grabbed my gear and ran in as fast as I could, which of course made me all sweaty right off the bat. Senior watched me with disappointment written all over his face as I rushed in with haste to where the group was waiting. He then proceeded to give me the evil death stare.

"I'm sorry," I wrinkled my nose and tried to look small.

"You know this is the second time you've been late." I resisted the urge to argue back and point out that *technically* last time I was only two minutes late, but for him, *practically* on time is already late. I, on the other hand, had picked up "southern time" at some point after having lived in Southern Maryland now for eighteen years.

Instead, I sucked up my pride. "I'm really sorry, Senior. It won't happen again." Excuses were irrelevant in the military and making them often only perpetuated the situation. I had learned that it was best to just suck it up and take responsibility for my actions. Eventually, this became a theme that I incorporated into my parenting skills and would attempt to drill home with my kids, as well. And I must say it was the weirdest feeling to feel like you are 'in trouble' when you're 41 years old. I was usually the discipliner.

"I hope not. Now get your stuff together, we're going to be assigned to our Humvees here soon."

As predicted, after an hour of waiting, we were finally told to line up to be assigned to our Humvee. Each Air Force guardsman was paired up with an Army guardsman. In typical Army fashion, they stood at attention in a neatly formed line. We waited for our name to be called out by our group leader, and one-by-one we casually lined up behind them in our own Air Force fashion. And no matter how much I begged and tried to bargain, my Chalk leader would not let me drive the Humvee, which I later realized was probably a wise decision, as driving them was not exactly like driving my SUV.

I have learned that an Army and an Air Force Captain is <u>not</u> the same as a Navy Captain. A Captain in the Army/Air Force is an O-3 rank verses an O-6 rank in the Navy. This is a very substantial difference. Every monthly drill in the Air Guards, we have our Commander's Call, in which the commander of the squadron (usually a Colonel) gathers us to give instructions, updates, etc. In the Navy we called them Captain's Call. Somehow, I cannot get Captain's Call out of my head and when I announce that we need to head out to Captain's Call, I get met with very confused faces. Thankfully, I have not yet said it in front of the Colonel.

So as the *Army Captain* belted out the instructions to the Army drivers in the front row, the young man adjacent to me caught my attention, as he had assumed a plank (push-up) position some time during the line-up. "What's up with him?" I whispered to the girl to my left in line.

"I was wondering the same thing. Is he *exercising* right now?"

I shrugged. "I mean, it is the Army. They do some weird things... On second thought, do you think someone *told* him to get like that? He looks pretty junior."

"Right now? Before we go out on a mission?"

"I know. It's not like we're in boot camp or anything." I knew the Air Force did not call it boot camp like the Navy did, but in that moment, I could not remember what they called it. "Does the Army do push-ups *after* boot camp?"

She just shrugged and we quickly turned our attention back to the Captain, as he gave us the stink-eye, I assumed for talking in ranks.

The Captain went on and on for a while about radio calls and our assignments. I tried to pay attention, but all I could do was watch this boy in front of me. As his arms began to shake and his back started to sway to the cement beneath him, I deducted that he was not doing this by choice. It was so distracting that I could not hear anything the Captain was saying. All I wanted to do was to just go over there and help him up. *Ughh*. I was so not good at maintaining military bearing. I am not sure how I survived this far.

Finally, after what felt like eons, the Captain was done speaking, and we were all dismissed to our vehicles. As I was walking out with my Army Humvee driver, he introduced himself as Jefferson. He could not have been much older than my daughter that just graduated high school. After I introduced myself, I asked him, "so what was going on with the guy next to you?"

"Oh, he got in trouble because he went to the Px (store) instead of coming directly here. He was late to get his Humvee prepped and lined up."

"Seriously? That's crazy! They still make you guys do push-ups?"

"Oh yeah. Well, I mean, he deserved it," Jefferson shrugged. As we walked out, I waved a goodbye to my disappointed Senior Sergeant who suddenly did not seem so bad. I mouthed the words, *'I'm sorry'* once again, and I thought about how thankful I was to be in the Air Force instead of the Army.

* * *

Riding in an armored Humvee was by far a different experience than riding in the LMTV. We were low to the ground and it was quite challenging to see out of the tiny rectangle, hazy-smudged glass windshield. The inside of the Humvee smelled damp and a little stale, so I slid my window open all the way; a full two inches was as far as it would open. That was going to make it challenging to wave to people. Due to all of the extra armor, there was a major blind spot on the right side and I had to clear that side any time we changed lanes or turned—which sounds easier than it really was due to the tiny, awkward hazy windows. Most of the time I was pretty good at it. There was only this one time that I underestimated our size and the side mirror hit a very large sign... but only one time. (For the record, both the Humvee and the sign were just fine.)

I had to remind myself the Humvee was built for wartime operations, not comfort cruising in the city. It supported four people and a bunch of gear, but it had a very claustrophobic vibe for me, as I had never been in one before. I am sure with time most people just get used to it, maybe even come to love it, as I had once come to love our dirty old plane, the mighty P-3 Orion when I was in the Navy.

We pulled up to the street as directed and parked behind the Metro Police car. I had to think it was a good sign that things were beginning to settle down, as for the first time, we only had one law enforcement official with us. I opened the heavy door and stepped outside of it to stretch my legs. The wind blew and once again I was greeted with the sweet scent of the magnolia tree blossoms. I could not get enough of it. There was just something about the smell of magnolias; it was the smell of contentment. It was like nature's way of saying everything was going to be okay.

I was about to introduce myself to the police officer on duty with us, when I was approached by a tall, blonde man. He was dressed in casual clothes, but as soon as he was in front of me, he flashed his badge at me. I did not really know which badges were legit, so I just nodded and pretended, as if I had to approve of people flipping their badges at me every day.

"Hi there, ma'am, how are you doing?"

"Oh hi. I'm good. What's going on?" I asked him.

"I'm with the Secret Service, and I just wanted to stop by and thank you guys for being here." My partner was stepping out of the driver's seat of the Humvee and made his way over to the sidewalk by me to listen.

"Oh, of course, well thank *you* for all that you are doing," I said back to him.

"No, really," he continued, looking back and forth between the two of us. "That first night, when the riots began and we stood in our formations, all I could hear from the crowd was them yelling, 'throw it at the tall guy! Throw it at the tall guy!' And they began throwing the bricks at us. But then you guys came out and stood in front of us with your shields." He paused for a moment. "Honestly, I don't know what would have happened if you guys hadn't come in that night. We were so outnumbered, and everything just kept getting crazier." His eyes looked tired, but sincere. Clearly, he had been through a lot.

My partner nodded and said, "I'm glad we could help."

"Whenever I see the military folks, I'm just trying to let them all know how appreciative we are." And with that, the tall guy nodded back and kept walking down the street.

*　*　*

I watched a small figure emerge from the police vehicle in front of us and walk in my direction. She had long, straight black hair and an infectious smile. Other than her short stature, she looked as if she had the beauty to have graced the fashion runways of Paris.

"Hi there," she called out in a friendly, bubbly song.

"Hello," I said with a smile. She was the first woman federal entity I had worked with outside of the National Guard.

"Hey, I just wanted to introduce myself. I'm Brooklyn. Are you guys doing okay? Do you have air-conditioning and water?"

"Oh yes," I answered. "We are good and well stocked up. It only took a few days, but they are actually supplying us with water now." I joked. And as most of my jokes go, I often forget that not everyone knows my sense of humor and the entire situation often just turns awkward.

But Brooklyn and I clicked. "Ha! Don't you love it?"

I laughed because I actually did. I hated it all and loved it.

"So where are you from?" She began. And within two minutes, I felt like I met a woman I had known my whole life. She was from Southern Maryland, near where I lived now. In fact, her mother lived there still and had been watching her two nine year-old twin boys, as she and her husband (who was also on the police force) had been working around the clock. They had been doing 12-hour shifts for three weeks straight.

"I'm so excited. I finally get my first day off tomorrow," she exclaimed. I'm going to get up at the crack of dawn, go into my mother's house and wake my boys up. They will be so mad and then so excited. It'll be great!" She laughed.

As the sun continued its journey through the sky, our conversations on the street continued. We would still glance around at various times, if there was a noise or an interesting passerby, as we were still alert to our duty. It was just so refreshing to have someone to converse with about the time we once felt 'normal.'

I cannot tell you what we were talking about at the exact moment he caught my eye. There was something about his look that made me forget what I was talking about all together. He passed slowly, but with intention. He carried a shopping bag in his hand. His eyes met mine as he passed, and I smiled, cautiously. And as I did it, I regretted my smile, as if I had just created an invitation.

I felt a moment of relief as he passed us, thinking he was gone. But what I did not know was that the moment was only beginning for him, as he thought better about passing up an opportunity.

What happened next felt as if I was watching myself in a movie. The tall, skinny black man turned back at us, his eyes narrowed, and he shouted, "you gonna lock up some niggers tonight!?"

We were both stunned to speak. There were no words.

And so, in our shocked silence he continued to go on, this time bending down towards our faces to be sure that we heard him. And he repeated louder, "I said, are you gonna lock up some niggers tonight?"

It was as if he was angry at us for not responding. But what kind of a response was he after? We were beyond any logical response by now. It was not a real question. It was an accusation. And I was so dumbfounded, I had no idea how to react. I cringed. I could not say anything, but my physical reaction lost all possible chance at being stoic.

"You heard me," he went on, getting closer. I glanced up at Brooklyn. She came up to his chest as he towered over us. But she drew on her training and handled it like a professional. Clearly, this was not her first time.

"Sir, that's enough. You need to keep walking." And she said it with such authority and assuredness, I was so relieved to be in her presence. Apparently, this is what she did. And the man must have been convinced as well, and he turned around and kept on his way, bag in his hand.

Thursday, June 11, 2020

It was the most absurd thing. I could not help but wonder, if Brooklyn and I had not been in our uniforms—just two moms on the sidewalk having a conversation about our kids, would he have spoken to us like that? What was it about being in uniforms that suddenly made us less human to him— what was it about the uniform that made him feel like he had permission to speak however he wanted to us?

* * *

As the day dragged on, I kept alternating between sitting in the Humvee and standing on the street corner, trying to determine which place was cooler. In the truck, it was shaded, and I could sit down, which took less effort and I felt like I finally stopped sweating. On the street corner, I was in the full breeze… that picked up about every few minutes or so and if I wished really hard, the wind blew, and it felt so amazing, like whimsical fairies blowing their breaths of minty icicles on my skin.

I chatted with my Humvee driver. He was eighteen and had just come off of orders for COVID Domestic Operations. In fact, he had picked up COVID—apparently as an asymptomatic carrier. I still was not entirely sure what that meant. He had it, but did not suffer from anything? Regardless, he had done all of the required quarantining. There was something about the magical fourteen-day isolation period that instantly assured the public that everything would be okay. So naturally, he was back on duty. At this point, with all of the people I had been around the past couple of weeks, I did not know how I could not have COVID.

* * *

I would watch the streetlight next to Humvee cycle through the colors. When it was green, all the traffic flowed, everyone was content, satisfied. After a while, it would switch to a quick yellow, followed by a red—a red that almost brought about an urgent sense of defeat. When it was red, time seemed to stand still, people were impatient, and restlessness grew.

In a moment that was like any other, a yellow jeep full of young guys pulled up next to us at the stoplight. I was standing outside of the Humvee, leaning on the open door at this time. Jefferson was sitting in the driver's seat. There was the usual music blasting coming from the jeep, however, it took me a minute to realize that they were not just carrying on loudly; they were targeting us with their shouts. They were made up of a multitude of individuals, some wearing their hats backwards, as if it was an outward demonstration of their values.

As I stood on the passenger side of the Humvee, next to the sidewalk, I could hear the shouts, but they were muffled and what they were all about did not register immediately. And then I realized that they were yelling at Jefferson. I looked over at him, and he had a dead stare straight ahead. His gaze did not falter. I thought about the possibility of them throwing stuff at us, and I decided to climb inside of the vehicle for safety.

The guys kept yelling, and the words were unspeakably ugly, calling him all sorts of offensive and brutal names. They were attacking Jefferson, as he was the nearest target. *Why did this red light seem to endlessly linger?*

Finally, after more ugly shouting and agonizing waiting, the light at last changed and the atrocities drove away in the jeep. And then it was silent.

"Are you okay?" I quickly asked Jefferson.

"I'm fine," he replied.

But I was not convinced. "Oh my god. They were awful. What is wrong with these people? I could only hear about half of what they were saying, but I know it wasn't pleasant."

"No," he laughed. "It wasn't pleasant."

"I'm so sorry." I did not even know what to say. "How are you so good at handling all of this? I am the worst! I don't know how we're supposed to act anymore."

"Really, it's fine." The eighteen year-old boy was telling me. "I've heard much worse in my life."

"You have? Ugh. I'm so sorry. No one deserves that. No one."

"Hey." The young man that was driving me around the streets of DC in an armored Humvee looked over at me with the wisdom of a thousand lives. "My grandma always told me, 'don't waste words on people who deserve your silence.' So, I don't."

"Wow. I kind of love that. She sounds very wise," I thought out loud.

"She really was."

***

Friday, June 12, 2020 — Another Day Off

Saturday, June 13, 2020

Armory/US Park Police Station

"Everyone be cautious when ordering any food while in uniform—especially if you're in the city," Intel briefed us that morning. "Apparently there was glass found baked into a pizza that was delivered to the South Carolina National Guard's hotel."

Of course they did. Why wouldn't there be glass baked into a pizza in D.C. for the military to consume? These crazy-ass stories were beginning to shock me less and less at this point. I was beginning to think that the people with nothing but hate in their bones were capable of anything. The more absurd, the better, as they would receive the attention that they sought. It was sad, really, as there were people out here that were advocating for a change for the good—and so much of this time they were being overshadowed by the opportunists and those that were just looking for destruction to further the divide for their own agenda.

I grabbed one of the boxed sandwich lunches that had been sitting out for a while on the food table and brought it back up to the bleachers with me. I opened up the sandwich to put mustard on it, and although I (thankfully) did not find any glass, I did find that the cheese on this 'served cold' ham and cheese sandwich had completely melted from the ambient temperatures of the Armory in the time it had been sitting out. I smeared extra mustard on it to overpower the soggy bread and ate it anyway, because there were not a lot of other options at that time. Plus, I had eaten much worse in life.

It was a slow day. We had troops on the streets, and we were hanging back at the Armory, waiting to either cover the night shift if needed or go in as reinforcements if needed. In summary, we were waiting to be needed. On this day, I had learned from my wasted hours of the past and brought my computer in to work while I waited. I was not sure how long I would be activated, as there was talk of extending us past the two weeks that we had almost completed. I was quickly running out of allotted time at my work, so any extra hours I could get in was helpful. Plus, as things were constantly changing at my job, I was just trying to keep up with emails and what was going on so I would not come back to thousands of outstanding emails without any situational awareness.

* * *

As the day passed into evening, it was more and more apparent that we could be going home early tonight, possibly even by 2100 or 2200 (9 or 10pm). And then the call came in. We were told we had 10 minutes to be in full riot gear and outside of the bus. Crap. That meant I had to run to my truck to stow my computer (I did not want to leave it on the Armory floor), and grab all of my gear. The parking lot was not exactly close, but I hustled and after some scurrying, we all ended up in full gear in front of the buses as directed.

Three buses of us loaded up and then waited to depart. And waited some more. Finally, Ripper came on board and said simply, "Hey guys, remember your training. Stick together and always be on the alert. Look out for your wingman." And even with the serious pep-talk, they still had no word for us on what exactly was going on.

Apparently, we were waiting for a police escort. As soon as three different police vehicles arrived with flashing lights and sirens, we followed them into the city. We (cautiously) ran lights as directed, and drove down all sorts of questionable alleyways, as if we were taking every backroad possible. Somewhere along East Basin Drive, we pulled over on the side of the road and waited. And waited some more. After an hour or so, the buses pulled into the US Park Police District One Station parking lot. We were told that we could get off the bus, but to not go far at all, as we would have to load up within two minutes if necessary. After sitting on the bus for over an hour with all of our gear, it felt nice to stretch our legs. It had rained and stopped and the wet night had brought a chill to the air. And so, in the most anticlimactic end to the night ever, we waited outside of the police station until we were finally given the all-clear at midnight and loaded back up to go home.

As the story of the night goes, the police were called up. Several people dialed 911 to ask for police assistance as fights had broken out in a "Defund the Police" protest. We were called in to be on the ready in case the police needed backup, as this was a protest in which the police were probably not welcomed, despite the fact that the police were called to assist.

It truly was difficult to wrap my head around, as it was one of the most ironic things I had ever heard of.

When we got back to the Armory, I slowly dropped off all of my riot gear and stowed it. I was tired. I climbed in my truck and was just ready to go home. It was sometime nearing 0200 (2am), when I passed through Southeast D.C. I was surprised to see any traffic this time of night, but it never seemed to cease. As I came to a stop at a red light, I glanced over at the car to my right, parked next to me. Inside the car were four men, probably in their twenties from what I could guess in the dim city lights. The driver looked over at me and did a head motion that was almost like he was looking me up and down, despite the fact we were sitting down.

I quickly turned my head forward to the red light, feeling uneasy. Once again, I was stuck behind a car. *I was in such a hurry to get home. Why didn't I take the extra five minutes to change out of my uniform?* Then I heard shouting coming from the car. My heart began to beat faster, I did not want to look over, but I didn't know what else to do. *Do I call someone? That way they would know where I was last if anything happened. Stop. You're overreacting. They are probably just talking loudly to themselves.* But I knew better.

And then the light turned green. *Oh, thank you, Jesus.* As the car in front of me pulled forward, I began to accelerate. But before I could go more than a few inches, the car next to me sped up and crossed over in front of me and completely stopped. I had nowhere that I could go. *Oh my god. What do I do?* I knew I had an old knife in my center console somewhere, but what good would that really do me against four grown men? Maybe it would buy me some time. *Do I run?* I had never been in a situation like this before. I had never been so afraid in my life.

And then, after what felt like an eternity, but in reality, was probably less than a minute, the longest minute of my life, the car squealed its tires and took off. I sat for a second, trying to gather myself, but then realized I needed to get the hell out of this city as fast as I could. I drove off, holding my breath until I was well out of there and onto familiar roads. And then I finally allowed all of my emotions that I had been holding in to come to the surface.

Sunday, June 14, 2020

The COVID

"Whatcha doing?"

"Well. I just woke up. And I've kind of been walking around in circles looking at my house. I'm not sure where to even begin." The kids' laundry baskets overflowed and there were wet towels hanging on every surface between their bedrooms and the bathrooms. The dishes in the dishwasher were clean, but the surface of the counter was lined with more dishes, random odds and ends were everywhere, the bottles that needed to be recycled, the junk mail was stacked high and in several piles across the table... and it went on...

"I thought I was going to have a day off, but turns out..." I lowered my voice. "It's a disaster. I know he tried, but he doesn't see the messes like I do. So, I'll be cleaning today." *I love my husband. I love my husband.* I kept repeating it in my head. *He has taken care of everything for me. Kids and animals are alive, and the house is standing. That is all that matters. Don't complain about the mess. Suck it up.* I kept telling myself these little affirmations. Maybe they were not exactly affirmations. Perhaps I was trying to talk myself out of losing it on him. There was, after all, quite a bit of stress built up in my head and body. And I knew in the end he was doing the best he could do.

"Aww. I can help! Let me come over!" My best friend of 30 some-ish years always knew just what I needed and could read between what I said, and what I really meant.

"No... you shouldn't. I need to get tested today and I'm supposed to quarantine for like a month or something."

"No. I'm coming over. I don't care anymore. I miss you so much and you need help. Please, I have hardly seen you for two weeks." I could hear the emotion in her voice. Usually she was the hypochondriac—er, cautious one in our relationship, so this was very out of character during the pandemic.

"Anna, I don't want to put you at risk. Listen. If anyone has it, it's me. I've had hundreds of people screaming right into my face every day. I've been around thousands of guardsmen, and there have been five confirmed cases among us now. Also, I've hung out with the hairy rats on the street. I'm pretty sure I have the COVID *and* street-rat disease. And whatever else you catch from the sidewalks of D.C. Hepatitis? Ringworm?"

"It doesn't matter. I need to see you. I'm coming over, so just stay put and I will help you," Anna insisted. She did not get bossy very often. That was usually my job. So, I listened, smartly. And it was so good for my soul.

This disease hysteria has brought on such a wave of emotions all around. It had been unlike anything that has ever happened—to have a pandemic when everyone has access to social media. Lots of shaming was going on, whether you stayed home, or if you went out. If you wore a mask, or if you did not. Everyone had a different opinion, and everyone wanted to share their opinion. The politicians were constantly changing their stories. It was ridiculous and no one really knew anymore what was fact, and what was fiction. When the riots broke out, the disease was no longer convenient to quarantine from, so we (the National Guard) were all ordered to control the crowds of thousands. And I can promise you there was no six feet of separation in many of those circumstances.

After a while, it became more and more apparent that no one truly knew what this was or what we should be doing for it, or how severe it really was. We are still trying to find the answers.

But one thing that we did know is that our personal mental health and our mental health as a whole country was suffering tremendously. The suicide counts were beginning to skyrocket. Alcoholism and drug abuse continued to climb at a tremendously fast rate. People need people, and during this time that had heightened levels of tumult, we *really* needed each other more than ever.

So that I could find out for sure and put my mind at ease (and after I cleaned my house, of course), I went into MedExpress that afternoon and had my test done. They called me the next day to tell me it was negative and I was just relieved that I did not have to do the call of shame, because after the previous two weeks, I really would have been hurting with the amount of people that I had been around.

Tuesday, June 16, 2020

Leaving the War

"Report to medical 0900 for medical eval. You are leaving the war," my Chalk leader texted me. Of course, it was written in humor, but part of me could not help but feel there was an ounce of truth to that statement. True, it could not even compare to the horrors of a 'traditional war.' (Is there such a thing as a "traditional war"?)

This was not a real war. However, various locations across the country had become a battleground for justice for nearly three weeks now. There were cities where smoke was still rising from the ashes. Countless businesses had been destroyed. People had lost their lives. People had sustained injuries. I do not think we were even certain who the enemy was, or convinced that the 'sides' really disagreed at all.

Perhaps that is a familiar feeling that creeps in during wartime.

* * *

I was nervous to go back to the city. I had not been there since my terrifying moments in Southeast a few days ago. It was early on a Tuesday morning, so on my way to the city, I pulled into the local Wawa gas station, as I knew their coffee was some of the best coffee around. I threw on my cover, grabbed my wallet, and made my way inside. Despite the warnings, I had never been nervous down here about wearing my fatigues in public. This area was small town and had military influence all around it, as the nearby base was one of the most common sources of employment for the locals. American flags flew in many yards and military alumni could be seen on many bumper stickers and car decals.

I filled my cup full of the heavenly brew, taking a moment to enjoy the happy scent of morning.

As I made my way to the checkout, my eye caught a glance of the woman that was working the register, and I felt my muscles reactively tighten up in defense. She was older than I was and had a cheerful energy as she worked behind the counter. But something caused me to hesitate, an apprehension that had been foreign to me three weeks ago. Her skin color was different from mine. And I was in uniform.

Holding my breath, I carefully placed my coffee on the counter in front of me and prepared myself for whatever that could come next. I placed two dollar bills in front of her.

"Oh, no, Honey. You don't pay for that," she looked at me and pushed my coffee back towards me. "Thank you for your service," the lady looked at me, and although we were both wearing masks, I could tell by her eyes that she was smiling.

Her words caught me by surprise, and instantly my eyes began to well up with emotion. "Oh, um. Thank you," I said quietly back to her and walked out of the store to my car, where I climbed in and just sat for a moment. I did not mean to, but I started to cry. Like hard, ugly cry. The cry that let everything out that I had been holding in for so long now. I was thankful. And I was ashamed. I was proud, and I was scared. I was confused and I was determined. Every single emotion possible seemed to run through me in that minute in the parking lot. Thank god for tinted windows, as I know I must have been a ridiculous sight.

I know this lady could not have possibly realized the effect she had on me that day by the simplest action. At a time when I felt broken down mentally, she showed me kindness. I thought about everything that I had experienced in the past weeks. How the actions of others impacted me so much, whether it was good or bad.

When at last every bit of emotion in me had come to the surface, I took a deep breath and slowly let it out. I had nothing left in me to give. It was time to check off of duty and go home.

From where I stood, the water looked infinite. Only a glowing gray sliver broke where the violet sky touched the earth. There was not a ripple of wind or any sign of movement on the water, giving it a mirror-like appearance. All was calm again. The turbulence and uncertainty of the rolling waves had subsided, leaving nothing but a reflection of whomever peered into the infinity pool as they passed.

The evening was still. The sky was bright. The riots were largely over. Those who sought to disrupt and cause damage for political aim had been outlasted, for now. After all the theatrics, painted words in the street, and ugliness, there remained an undeniable faith that this nation still has a solid spine of good and honorable people with patriotic intent. For the moment the water was calm, without waves. It reflected the faces of fellow Guard members, the law enforcement teams, special agents, and task force members. It reflected the peaceful protesters who had the courage and conviction to come out, despite the pandemic, and have their voices be heard.

Standing on this ground, I was completely surrounded by so many brave Americans that all came before me to shape this country, to make it better for their children, and generations after. To my left were the 58,320 names of those that gave their life in Vietnam. To my right was Dr. Martin Luther King Jr. In front of me was the memorial to honor the Greatest Generation, those veterans of World War II. And behind me sat Abraham Lincoln, a man who changed the course of America for the better, probably more than any other single person. Each one of these people, and so very many more, put one foot in front of the other, and served the people of their country. The sacrifice and commitment to a cause that one truly believes in lasts longer than a passing wave and should have more permanence than an angry mob.

None of this would exist without courage and dedication to a cause greater than oneself. People with conviction, people that are not afraid to stand up, and not afraid to speak up for what they believe in are the ones that inspire those who need it most. These are the people that shape our country—constantly molding it, and reinventing to improve ideals, morals, and values until it becomes our future. The same energy and enthusiasm that draws people to the street needs to be harnessed and directed to create a platform for positive and long-lasting change. The fringes will always be there, but the fabric of our nation believes in freedom, in liberty and justice for all.

I leaned over the edge of the pool. My forehead was still bruised, and I had dark circles under my eyes. They were tired. I still could not make sense of everything that had happened. I only hoped that what we did was for the greater good. I hoped I was on the good side of history, someone that helped shape it into something better.

America will never be a perfect country. There will always be injustices as life continues to evolve. Nothing is perfect. But that is the beauty of it—

and I would take an imperfect country where I have the freedoms to stand up for what I believe to be just, or to defend those that are there to make a difference. And those people, the ones that have the courage to make the changes happen, are what makes this country perfect, despite its flaws.

    As I turned to walk away from the pool, a gentle, refreshing breeze broke the stillness of the night. And I could smell the magnolias.

Epilogue

"Hey there's Mike!" I glanced over to see this tall figure making his way up the hill towards where we were sitting at the picnic table under a pavilion. From where we were sitting, we could watch the ducks swimming along in the small lake out in front of us, diving their heads underwater every so often for food. It was a beautiful autumn day. The weather was just beginning to cool down and the humidity was finally giving us a break from its pain.

It was our first 'morale' event all year, due to COVID-19; the disease that shut down everything and anything fun. We planned this picnic and outdoor games in the afternoon of our Saturday drill. It was time we did something relaxing as a squadron.

Not everyone was able to make it, but as I looked around, it was such a happy sight to see our people with smiles on their faces as they caught up with each other over barbeque. Most were in civilian clothes, as it had been authorized for this afternoon. The ladies wore their hair down, and the men wore their favorite sports team t-shirts. It was hard to imagine just three months ago we were in bulletproof vests and helmets.

As soon as Mike saw us, he smiled. "Hey there!" he called out.

"Hey! Thanks for holding down the fort so we could get this all set up. Did all the planes get back on deck okay without issues?" MSgt asked him.

"Oh yes, yes. All is good in the hood." It was impossible for me to imagine

Mike without his cheerful demeanor. I just had never seen him any other way but smiling—even on the longest days. And I know it could not have been easy being the only guy in our Airfield Management shop.

"Well there is still plenty of food, so go fill up a plate before we start cleaning up."

"Alrighty. Oh, and Pyrah," he called out to me, "just wanted to let you know, I'm going to send you something today."

"Oh good. That will be great."

MSgt sighed and smiled. "I will too. I've been meaning to, but every time I sit down at my computer, I just can't think of anything to write."

"Anything is great. Just start telling me a story about something you saw there."

"Okay. I'll work on it tonight. I promise," she said with a reluctant smile.

"Well thank you." I smiled back at her and looked over my right shoulder to see Big Daddy headed towards the parking lot. "Ooh, I have to go catch Miles before he leaves. Be right back to help clean up."

"Hey!" I called out as he was walking away. "Don't avoid me," I said with a smile.

"I'm not avoiding you."

"Uh huh. So anyways. Are you really going to do it?"

He looked up at the sky dramatically, as if I was asking him to cut off his left hand. "Yeah, I actually already started."

"Ooh you haven't given her your story yet? Dude. I did." Weather smiled, pleased with himself.

"Hey, I've been out in the field. And I don't have a ton of stuff to say, really."

"Oh my gosh. You have so much to say. Alek's was so great. You guys were the ones right in front of the White House. You were the ones that got hit with bottles of unspeakable things. We need to tell your story." I was relentless.

"Okay fine, I will get you something tonight. I promise."

"What are you guys talking about?" I turned to see Louis come walking up with M following right behind him.

"She's writing a story about all the crap in DC last June," Alek answered him.

"Oh yeah?"

Epilogue

"Yep. Want to help out too?" I asked them. "Do you have some stories? I'm just trying to paint the picture of what it was really like down there, because so many people are trying to decipher what's real and what's not."

"That sounds awesome. Yeah, I have some stories. I'll definitely send you something."

"Yeah, me too."

"Perfect! I think this will be pretty cool," I said with a cautiously optimistic smile.

A Note from the Author

None of the stories in this book reflect any official policies or reports from the National Guard, or any branch of the military, federal entities, or its members.

All of the stories are strictly personal experiences and opinions of the authors and contributors. Most names have been changed.

About the Author

Julia A. Maki Pyrah grew up in Deer River, Minnesota, and enlisted in the Navy as an Aviation Warfare Systems Operator aboard P-3Cs. After the Navy, Julia settled in Maryland with her husband where they both ended up working for the Department of Defense as civilians. They have three children together, which initially inspired her to write children's books about the military.

Julia now works in Airfield Management in the D.C. Air National Guard. When she is not writing, raising a family, or working, she finds her passion in volunteering for organizations supporting veterans. Other titles she has published are: *My Mom Hunts Submarines; All Hands on Deck! Dad's Coming Home; Still My Dad; From the Sky;* and *What They Don't Teach You in Deer River.*

About the Co-Author

Nicholas A. Cotroneo grew up in Eagan, Minnesota, and attended high school at Saint Thomas Academy. He later graduated from University of Notre Dame and was commissioned as an Ensign in the United States Navy. After earning his Naval Flight Officer "Wings of Gold" in Pensacola, Florida, he went on to serve in Operation Iraqi Freedom—earning 6 Air Medals along the way.

He continues to serve his country as a Defense Contractor, and finds himself fulfilling his duties as husband to his lovely and understanding wife Mia. He is also a father to three wonderfully inspiring daughters. While his family keeps him busy, he has been inspired to find time to write by a close friend and colleague.

About the Publisher, Tactical 16

Tactical 16 Publishing is an unconventional publisher that understands the therapeutic value inherent in writing. We help veterans, first responders, and their families and friends to tell their stories using their words.

We are on a mission to capture the history of America's heroes: stories about sacrifices during chaos, humor amid tragedy, and victories learned from experiences not readily recreated—real stories from real people.

Tactical16 has published books in leadership, business, fiction, and children's genres. We produce all types of works, from self-help to memoirs that preserve unique stories not yet told.

You don't have to be a polished author to join our ranks. If you can write with passion and be unapologetic, we want to talk. Go to Tactical16.com to contact us and to learn more.

www.ingramcontent.com/pod-product-compliance
Lightning Source LLC
Chambersburg PA
CBHW041127110526
44592CB00020B/2719